Alcohol, Drugs and Driving

Modern Problems of Pharmacopsychiatry

Vol. 11

Series Editors
Th.A. Ban, Nashville, Tenn.
F.A. Freyhan, Washington, D.C.
P. Pichot, Paris
W. Pöldinger, Wil/St. Gallen

S. Karger · Basel · München · Paris · London · New York · Sydney

Satellite Symposium of the 6th International Congress of Pharmacology on Alcohol, Drugs and Driving, Helsinki, July 26–27, 1975

Alcohol, Drugs and Driving

Volume Editor
M. Mattila, Helsinki

With 22 figures and 17 tables, 1976

S. Karger · Basel · München · Paris · London · New York · Sydney

Modern Problems of Pharmacopsychiatry

Vol. 1–5: Details on request
Vol. 6: The Role of Drugs in Community Psychiatry
VI + 128 p., 3 fig., 3 tab., 1971. ISBN 3–8055–1200–7
Vol. 7: Psychological Measurements in Psychopharmacology
VI + 267 p., 25 fig., 54 tab., 1974. ISBN 3–8055–1630–4
Vol. 8: Psychotropic Drugs and the Human EEG
X + 377 p., 95 fig., 19 tab., 1974. ISBN 3–8055–1419–0
Vol. 9: Trazodone
VI.+ 210 p., 86 fig., 83 tab., 1974. ISBN 3–8055–1781–1
Vol. 10: Genetics and Psychopharmacology
VIII + 132 p., 15 fig., 25 tab., 1975. ISBN 3–8055–2117–0

Cataloging in Publication
Satellite Symposium on Alcohol, Drugs and Driving, Helsinki, 1975
Alcohol, drugs and driving: satellite symposium of the 6th International Congress of Pharmacology on Alcohol, Drugs and Driving, Helsinki, July 26–27, 1975/volume editor, M. Mattila. –– Basel; New York: Karger, 1976
(Modern problems of pharmacopsychiatry; v. 11)
1. Accidents, Traffic – prevention & control – congresses 2. Alcohol Drinking – congresses 3. Automobile Driving – congresses 4. Drugs – adverse effects – congresses I. Mattila, M.J., ed. II. International Congress of Pharmacology, 6th, Helsinki, 1975 III. Title IV. Series
W1 MO168P v.11/WA 275 S253a 1975
ISBN 3–8055–2349–1

Printed in Switzerland by Thür AG Offsetdruck, Pratteln
ISBN 3–8055–2349–1

Contents

Opening Remarks

Ladies and Gentlemen

We all are aware of the great importance of the problem of the traffic safety, and so do probably the citizens all over the industrialized countries. As soon as the causes of traffic accidents as well as the countermeasures to control them are discussed, the opinions may differ even sharply and the whole issue readily becomes a political one. In Finland we have a governmental traffic committee, the members of which represent different political parties as they are represented in our Parliament, and a similar situation may exist in other countries as well. That kind of committee has a hard task to improve the quality, safety and availability of traffic, and there are many major activities which no doubt belong to a governmental committee and not to scientists.

The amount of traffic is an important constituent of its safety, and governmental bodies can control it by adjusting, e.g. fares on railways and on roads, by issuing special traffic rules, and readily by pushing up the prices of cars by heavy taxation. The last-mentioned measure has been practised in our country several times but the efficacy of that measure is difficult to assess. Driving a car easily gives such a convenience and pleasure which are difficult to outweigh. One measurable effect has been the dominance of small and light cars which in a way can diminish the traffic safety. *The driving conditions* are another important safety factor, and the proper supervision of the quality of roads and vehicles, illumination, etc. can no doubt influence the results, yet the costs may limit the progress along that line.

The quality of the drivers is an entity with many facets, some of which we can probably control to a varying extent by proper measures. Whenever a wrong and dangerous *attitude* towards the rights and welfare of other motorists and pedestrians is evident, the views for improvement are not very promising: like good advice for public health in general, information may not 'get in', and adequate supervision by police may prove too expensive. Inadequate *skills* may prove easier to correct by proper schooling and training. Dr. *Häkkinen* will present in his lecture how it is possible to predict an individual accident proness, and this in turn should be handled with extra training or/and with some restrictions. Whenever the driving skills and/or attitude are temporarily lowered by tiredness, age, disease, or by agents such as alcohol and drugs, they can be controlled by legislation provided that those legal measures lie on the solid base of research work, both epidemiological and in laboratory. This kind of research work poses several problems of both actual performance and interpretation, which have recently been discussed by *Silverstone* (Br. J. clin. Pharmac. *1:* 180, 1974) and which we have come here to discuss further.

While I apologize for the failure of some invited speakers to come here I look forward to gathering new data, ideas, and guidelines for a better approach to the problem of traffic safety. Welcome to the symposium!

M.J. Mattila, Helsinki

The symposium has been sponsored by Oy ALKO Ab, Finnish Medical Association, Hoffmann-La Roche Ltd., Liikenneturva ry. (Central Organization of the Safety of Traffic, Finland), Sandoz Oy, and Sigird Jusélius Foundation.

Mod. Probl. Pharmacopsych., vol. 11, pp. 1–10 (Karger, Basel 1976)

Efficacy of Law Enforcement Procedures

Concerning Alcohol, Drugs, and Driving

Robert F. Borkenstein

Center for Studies of Law in Action, Indiana University, Bloomington, Ind.

Until a dozen years ago, motor transportation by automobile was very inequitably distributed even among the highly developed nations. Three fourths of all automobiles, vans, and trucks and half the improved highways of the entire world were in North America (8). This system served only 7 % of the world's population. Because of this early concentration, the problems and dysfunctions of the system showed up sooner and methods of relieving them were tried many years ago.

Professional organizations such as the American Medical Association and the American Bar Association were active in these matters 40 years ago. Sweden recognized the alcohol factor in highway safety and was very active in studying it at least that long ago. It is for these reasons that much of the material I present will be of North American and Swedish origin. However, much of the information has been gathered from organizations such as the International Road Federation, the Organisation for Economic Cooperation and Development, and personal communications.

Dependency on the motorcar is expanding everywhere, even though attempts to substitute mass transportation have been accelerating due to the energy crisis. Today, among the OECD nations the highest vehicle density[1] is 55 in the USA and the lowest is 18 in Japan. Finland is at 29, close to the average – 32 (7). Ten years ago the range was much greater and the average lower. The problems encountered by the USA 40 years ago are the problems being faced by those nations with currently exploding motor vehicle density. While there are cultural differences, most factors and problems fall into amazingly consistent patterns. The subject of this paper is the role of law enforcement in relieving one of the principal causes of dysfunctions in the traffic system – the alcoholically impaired driver.

[1] Thousands of vehicles per 100,000 total population 1970 (OECD, 1974).

About 40 years ago an American jurist studied the role of the law, police, and courts in the prevention of automobile crashes (10). He listed five classes of offenses which were, in his opinion, highly correlated with traffic crashes. These were excessive speed, alcohol, improper lane usage, following too closely, and license violations. These crude cause-effect relationships comprise, even today, the basis of traffic laws, their enforcement, and subsequent adjudication and sanctioning. In spite of their crudity these cause-effect relationships have not been without merit. The death rate on American highways is the lowest in the world. Three years ago the rate was 3.3 per 100 million vehicle kilometers; 2 years ago it was 2.85, and last year it was 2.12. Figure 1 compares many nations (5).

While relatively unplanned approaches were used, opportunities for improvement were so great that even these were destined to have some success. They included better highways, driver improvement, and safety features on cars. Less than half a century ago most cars still had two-wheel mechanical brakes, plate glass windshields, hand-operated windshield wipers, and very unsafe tires. In most places, drivers were unlicensed. When one examines the results of these efforts in figure 1, it becomes obvious that the USA is now dealing with a very stubborn residue, and residual problems are bound to be more difficult to cope with than those which, by their very nature, seem to fade when crude countermeasures are applied. However successful the historical countermeasures have been, Americans cannot be smug about their 'success' because of the enormous size of the highway transportation system and, therefore, the sheer number of people being killed, even at this low rate. Figure 2 indicates motor vehicle deaths per million population per year in the same nations shown in figure 1[2]. Since motor vehicle deaths principally involve youth, their actuarial importance becomes even greater.

Bertrand Russell excused the use of crude cause-effect relationships *in the infancy of a science* but deplored it when means became available to identify intermediate factors. When one examines the bars in figure 1 there is no way of knowing what factors were responsible for what percent of the fatalities. However, we do know from improved information provided during the last decade that alcohol is involved in about half of traffic fatalities in the USA (10) and about one fourth in Finland (6). Yet the alcohol-involved death rate is about the same for both nations because these are percentages of different total highway death rates. From the crude figures available from a number of nations, alcohol involvement in highway deaths appears to reflect the cultural use of alcohol rather than effectiveness of countermeasures such as law enforcement. Alcohol-related crashes tend to cluster at the top of the crash-severity scale, both in terms of personal injury and of property damage (1). This is the basis of the high

[2] Ratio of highway deaths to total population (*Borkenstein,* 1974).

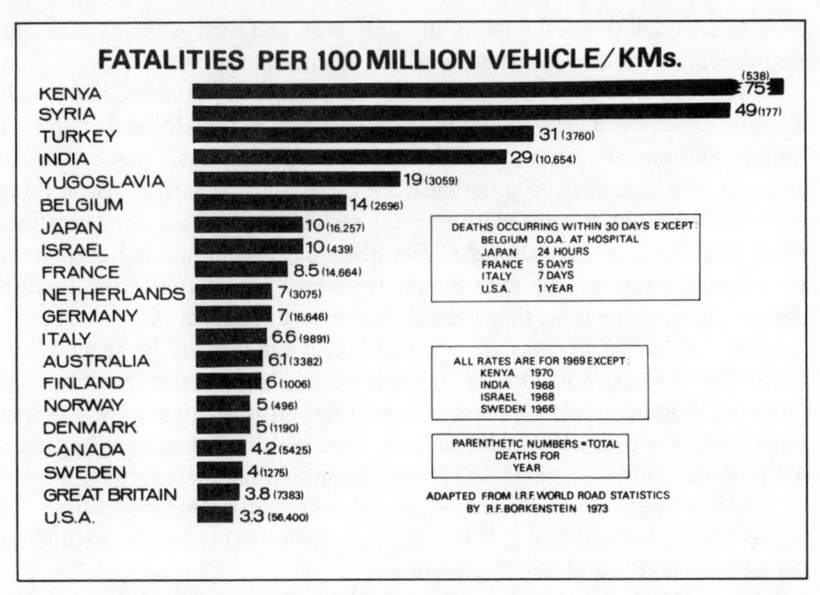

Fig. 1. Incidence of fatal road accidents in relation to the amount of traffic in different countries.

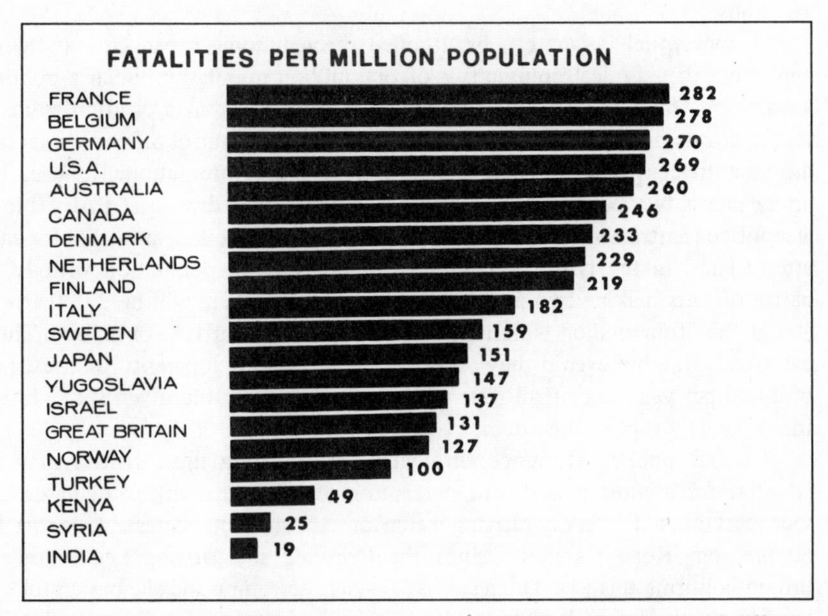

Fig. 2. Incidence of fatal road accidents in relation to the population of different countries.

priority accorded to alcohol factor in safety programs such as the Alcohol Safety Action Projects of the US Department of Transportation.

From evidence collected since 1940, admittedly crude but indicative, there is every reason to believe that programs directed at drinking driving have had little impact on the frequency of alcohol involvement in fatal crashes. Improved and generally excellent information from these Alcohol Safety Action Projects confirms the stubbornness of this factor in highway safety, and has provided much food for objective thought. The planning, execution, and evaluation of these projects provide the springboard for where we go from here. I will first discuss where we are now, then suggest some new directions.

First of all, let us consider what resources are actually being allocated to traffic law enforcement. On the average, 10 % of police resources are so allocated (7). Competition for police and court resources is fierce. Except in special traffic law enforcement agencies such as those which exist in New Zealand and some of the American states which have their own highway patrols, the police have never been very enthusiastic about their traffic responsibilities. Public demand to control criminal activities has almost always resulted in placing traffic law enforcement in an inferior position.

Let us briefly consider how traffic laws are being enforced, using the term 'law enforcement' to imply the entire system of apprehension by the police, adjudication by the courts or some alternative institution, and the imposing of sanctions.

A conceptual insight can be attained through some broad cuts at figures applicable to a typical community of one million population. Such a political subdivision will have available one patrol police officer in a position to make traffic arrests per 1,000 residents. On the average, each patrol officer arrests two drinking drivers per year. These shocking figures hold internationally. When the arrest rate is two per year, the frequency of violations, driving at BACs (blood alcohol concentration; 1 ‰ = 1 mg/ml) or higher will be at least 2,000 for each arrest. Thus, in the typical community of one million population with 1,000 patrol officers making two arrests per man per year, there will be 2,000 arrests and at least four million violations (driving with BACs of 1 ‰ or higher). These are broad cuts, but even if they err 50 %, the problem is apparent. An arrest rate of 2,000 per year to control four million violations is futile. Even if this rate is tripled or quadrupled, the attempt is still feeble.

General public deterrence through propagandizing these figures may be effective for a short period, but perception of low actual risk of being caught soon sets in and drinking/driving behavior is bound to regress. A report by *Codling* on 'Road Casualties since the Drinking and Driving Legislation' in Britain confirms this (4). The result is low-yield screening and the basic problem remains untouched, at least in terms of goal achievement – significant reduction of traffic crashes and fatalities and identification of problem drinkers.

A survey made a few years ago (2) showed that in a sample of 1,000 subjects from social strata of persons who ought to have been aware of and influenced by traffic laws, not a single subject was fearful of being apprehended by the police when driving home from a party after drinking too much. This perception of low risk of being apprehended is hypothetically the reason why drinking driving remains a stubborn residual problem.

The big question is, can we afford the resources necessary to achieve general deterrence and at the same time deal effectively with the individual driver? If emphasis on therapeutic measures is to be one of our principal goals, it means that we should be screening the corporate whole of the driving population to identify an optimal number of drinking drivers, to classify them as to their drinking problems, and to assign them to the proper sanctioning countermeasures.

Some time ago I interviewed police officers in several nations, asking them how many alcohol-related traffic arrests they had made during the prior year. The answers ranged from none to three. I also asked each one how many motorists he had stopped for hazardous moving traffic violations whom he thought should have been tested for BAC and possibly put through the formal system. The answers ranged from 75 to 100.

The reasons for this exercise of police discretion are numerous – police cynicism as to what support the charge of driving under the influence of alcohol will receive in the courts, empathy with the driver, the feeling that sanctions are far too severe, the feeling that the driver will be deprived of a substantial part of his livelihood if he loses his driver's license, the cumbersome system employed to handle the drinking driver through the charging, arraignment, and court procedures – these are among the many reasons given. At any rate, I have tried to set out some of the problems faced by those enforcing the drinking/driving laws. This is today. Now, where do we go from here?

In my opinion, our thinking has been unsystematic. Our approach has been largely catching a few fish from the sea of drunken drivers and making horrible examples of them. This approach apparently has not been successful.

Three quantitative terms can be used to describe the level of our activity directed at the drinking driver: maximization, optimization, and minimization. Maximization is not practical. It would mean apprehending every drinking driver under every circumstance and subjecting him to a tailor-made set of sanctions, ranging from punitive to therapeutic.

From the figures I have cited, it is obvious that even in our most ambitious programs we are applying enforcement, adjudication, and sanctioning at a minimal level. The question is, what is optimization? Optimization is the level of activity which, coupled with adequate and effective public information, will create a high degree of general deterrence consistent with predetermined goals. We can hypothesize on the basis of experience in other fields of endeavor that there is a critical point where optimization will be achieved. However, even when this optimal level is reached, it may not be a permanent condition. When the British so-called Breathalyzer Law was enacted and put into action, there was an immediate impact. We might say that the level of enforcement, even though very low, coupled with an excellent public information program, was optimal *at the time.* However, the driving public soon habituated to this situation and regression set in. So what is optimal at one moment may not be optimal at the next. We rebel at inflationary prices but soon acquiesce and habituate to them. Thus, optimization is not necessarily a fixed factor.

Possibly, instead of applying a constant level of effort against a factor such as the drinking driver, it would be more effective to use the principle of occasional emphasis. Enforcement would achieve the optimal results for some period of time. The effect could be measured by counting the frequency of occurrence of excessive BACs in drivers in the population at risk by employing a form of roadside survey. When the effect of enforcement would become quantitatively evident and consistent with predetermined goals, enforcement could be relaxed until a point of undesirable regression would be reached, when another emphasis period could be imposed until the desired suppression of frequency of occurrence of excessive BACs would again be evident. This would be a continuous conditioning process. A massive amount of the resources allocated to traffic law enforcement could be used during the reemphasis periods. Such a program would also release a large segment of those resources for other enforcement duties between the reemphasis periods, offering task variety for job enrichment to those enforcing the law. It is hypothetically possible that after a time, the life-style of drivers regarding their combining of drinking and driving would be permanently altered.

Our current traffic law systems are more than 40 years old. They were conceived when transportation technology was in its infancy. They were conceived before the behavioral and social sciences became involved in the field.

As in any profession, it is imperative to reexamine the tools being used from time to time to see whether they comprise the best of all worlds at the moment. If they do not, they must change. The principal tool in the current system of attacking the drinking-driver problem is the law. Let us take a look at what the law actually says. It is a crime (misdemeanor, infraction) to drive (be in control of) a motor vehicle while impaired by (under the influence of, intoxicated by) alcohol, or to have a BAC of, say, 0.8 ‰ or higher. There is no mention of deviant or bad driving. If we wish to use the term 'cause', only actions under certain conditions can be causal. The state of the driver can incite or aggravate dangerous situations. Thus the laws as they are today infer that combining drinking and driving is a crime, even though deviant or bad driving is not evident. Possibly the ambivalent attitude of the driving public is based in a conflict between perceived risk of crashing and risk statistics. Drivers at BACs of one promille and higher appear in crashes six times as often as *an equal number* of drivers at BAC zero (1). These are *relative* figures; however, the drivers in this group comprise only about 80 out of 10,000 (US figures) drivers on the road (3). Thus the *absolute* number of alcohol-related crashes which can be attributed to a random group of, say, 10,000 drivers, is low. The public is confused by this seeming clash of numbers and tends to prefer to view it on the *absolute* basis. Drinking driving is such a common phenomenon that it becomes 'bad' in the common mind only when it involves dangerous driving or results in a crash. Thus a possible powerful legal concept might be based on the premise that the presence of excessive alcohol in the driver should be evident in his mode of driving or in the fact that he has become involved in a crash. In practice, few drinking drivers are identified by other means, so this concept is a formalization of an existent informal system. In each objectively describable hazardous moving traffic violation, alcohol, if present in excess of 0.5 ‰ might be considered an 'aggravating factor'. Thus if a police officer, upon stopping a motorist for a hazardous moving traffic violation, has reasonable cause to believe the driver *has alcohol in his body,* he could arrest or cite the driver for the specific moving traffic violation *aggravated by alcohol.*

This would tend to remove the moralistic overtone of laws simply prohibiting drinking and driving and would tend to tie alcohol to dangerous driving behavior and thus make such laws and their enforcement more acceptable to the driving public.

Under this system, every detected moving traffic violation would become a potential alcohol-related offense if the police officer had reason to believe the driver had alcohol in his body. The subsequent process of handling such arrests must be streamlined. Arrests or citations for alcohol-related offenses should not increase by just two or three times but by, say, 10 or 20 times.

The thought of such an increase in arrests immediately suggests enormous increases in necessary resources. This need not be the case. In a typical American city of one million population, 325,000 hazardous moving traffic arrests are made each year. Of these, only about 1 % are reported as alcohol-related. If patrols were assigned when and where alcohol-related offenses and crashes tend to cluster, the actual number of *alcohol-related* arrests could increase dramatically. It might mean neglecting some other violations, but if alcohol is the highest priority factor, would not this be a logical decision?

If each of the 1,000 officers could be motivated to include alcohol as an element of the offense in only one case each 2 weeks, the result would be a phenomenally increased screening-in of dangerous drivers who have been drinking. The purpose of this drastic increase would be to create a general deterrence by coupling this program to public information.

A conviction might result in a standard fine, possible a percentage of the offender's annual net income, plus required attendance at a driver improvement–alcohol information school. This course would be available to *first offenders only.* By keeping sanctions for first offenders at a minimum level of intensity commensurate with inhibiting recidivism in drivers whose drinking is under control, the rate of uncontested charges would be very high – well in excess of 90 %. As part of the driver improvement–alcohol information school, a cursory test might be administered in an effort to determine the nature of the driver's alcohol problem. Sanctions, to be most effective, should be imposed as soon as possible after the law violation has been committed, to minimize opportunity for rationalization of the act.

Let us conceptualize how this system would work. I know that systems vary from nation to nation, but as an example let us theorize that, in a city of one million population, each of two full-time courts can hear 25 DUI cases a week. The two courts could try 2,500 cases a year. These would be only contested cases and repeat offenders. Thus, if 90 % of the cited drinking driving cases are uncontested because of reasonable sanctions and the two courts would handle 25,000 cases per year, this system could handle 25,000 citations per year. Since the recidivism rate is about 3 % per year, 750 drivers in this group of 25,000 drivers would be identified for more intensive sanctions.

To summarize, this paper is based on the following axioms: (1) Even with 'soaring arrests' and 'fourfold increases', current enforcement programs, however energetic, are not scratching the surface when measured against the existing problem. (2) Law enforcement must be optimized to modify the drinking behavior of drivers by putting teeth in the messages of the media. (3) The police and justice system which enforces most of the laws dealing with the drinking driver must be systematized to handle a problem of epidemic proportions. (4) All operations against the drinking driver should be primarily directed at general deterrence, and secondarily at apprehension of particular individuals, although in practice the primary goal may be achieved by vigorously pursuing the secondary goal. (5) The laws dealing with the drinking driver should be rethought and restated embracing current social science knowledge and legislative and legal trends. (6) Alcohol in the driver should be considered an aggravating factor in traffic offenses rather than a primary cause. (7) Sanctions should range from punitive to therapeutic according to individual need. (8) A time-series study of statistically significant proportions should be undertaken to provide a means of sensing change brought about by the countermeasures.

Perhaps we should be looking at the deterrent effect through optimal enforcement as our only primary means of reducing driving while impaired by alcohol, and at the same time recognizing it as a means of identifying problem drinkers who are not only problems in highway safety, but problems in their homes, on their jobs, and in other areas of their lives. It is possibly the only process by which we can obtain legal coercive control of such individuals in order to offer them appropriate treatment. This gives meaning to the task. If effective feedback can be provided to patrol police officers, to the courts, and to the supervisors of sanctions, this would provide increased job satisfaction. Evidence of reduction of frequency of alcoholic drivers in the driving public at risk, measured by surveys, would give meaning to the activities of the patrol officer. The principle of task variety would enrich the police job by reducing monotony. There is nothing more boring than routine police patrol.

To my knowledge such systematic usage of police and court resources has never been employed. All this requires a change in thinking. We must be concerned about the individual driver and at the same time about the public safety aspects as they affect the entire community.

This is a new methodology for a field that requires flexibility in thought. Introducing change into law enforcement and the traditional courts is difficult, but we must always be aware of the implications of an old Chimese proverb: 'The willow that bends is stronger than the oak which resists.'

Summary

Fatality rates in autobomile crashes vary widely from nation to nation. For instance, in 1970 Japan reported a rate of 11.4 and the United States 2.6. Finland stood about midway with 6.2 fatalities per 100 million vehicle kilometers 1970 (OECD, 1974). These rates reflect all fatal crash causes. The alcohol rates are quite imprecise for many nations but a comparison of the Finnish and United States figures reveals some interesting facts. The United States rate of alcohol involvement in highway deaths is about 50 % (of a rate of 2.6). The Finnish rate of alcohol involvement is 27 % (of a rate of 6.2). Thus the alcohol rate for Finland is about 1.7, and for the United States 1.3. These percentages provide a guide for enforcement officials. The higher the percentage, the higher the priority. Moreover, the role of traffic deaths in overall mortality is important. For instance, in Japan each year 25 traffic deaths occur in each 100,000 of population, while in the United Kingdom the figure is 13 (1970 figures; OECD, 1974). Another factor is available countermeasures. Safety features on cars, quality of roads, driver training, and cultural attitudes toward cars all contribute. Driver behavior controlled by law enforcement is a very important factor. There is evidence that enforcement of drunken driving laws is extremely lax even in those nations claiming vigorous programs. An enforcement program that will screen the corporate whole of the driving population on an optimal basis will identify those drivers whose drinking is a problem for treatment ranging from punitive to therapeutic on the basis of their needs and at the same time bring about a general deterrence because of a perception of the high risk of being apprehended.

References

1 *Borkenstein, R.F.; Crowther, R.F.; Shumate, R.P.; Ziel, W.B., and Zylman, R.:* The role of the drinking driver in traffic accidents. Department of Police Administration, Indiana University (1964).
2 *Borkenstein, R.F. and Klette, H.:* The perception of DWI laws. US Department of Transportation (1971).
3 *Borkenstein, R.F.:* A study of the frequency and characteristics of drinking drivers in a typical county. Department of Police Administration, Indiana University (1967).
4 *Codling, P.J.:* Road casualties since the 'drinking driving' legislation. Accident Investigation Division, Safety Department, Transport and Road Research Laboratory, Crowthorne, Berkshire (1975).
5 International Road Federation: World road statistics, Geneva, Switzerland and Washington, D.C. (1972).
6 *Linnoila, M.K.:* Personal commun. (1975).
7 Organisation for Economic Co-operation and Development: Research on traffic law enforcement: effects of the enforcement of legislation on road user behavior and traffic accidents (1974).
8 *Owen, W.:* Strategy for mobility. The Brookings Institution (1964).
9 US Department of Transportation: Alcohol and highway safety. A report to the congress from the secretary of transportation (1968).
10 *Warren, G.:* The traffic courts. National conference of judicial councils (1942).

Prof. *R.F. Borkenstein,* Center for Studies of Law in Action, Indiana University, Scyamore Hall 302, *Bloomington, IN 47401* (USA)

Mod. Probl. Pharmacopsych., vol. 11, pp. 11–21 (Karger, Basel 1976)

Characteristics of Driving in Relation to the Drug and Alcohol Use of Finnish Outpatients[1]

M. Mäki and M. Linnoila

Department of Research Liikenneturva, and Department of Pharmacology, University of Helsinki, Helsinki

Early questionnaire studies have demonstrated a moderate increase in the road traffic accident rate among drug users (4), without any particular emphasis on specific drugs or specific groups of patients. The only drug-specific epidemiological study conducted so far, demonstrated diazepam present in the blood of traffic accident victims admitted to hospital in 18 % of the cases. This figure was higher than expected from the consumption statistics (1). As the study was concerned only with those accidents resulting in injuries severe enough to warrant hospitalization, a majority of road traffic accidents were excluded from the data. The present study was conducted among three specific patient groups generally receiving drugs for prolonged periods of time. The issues of interest were: (1) the driving habits of chronically ill outpatients; (2) the use of drugs and alcohol among outpatients; (3) the relation between alcohol and drug use, and driving.

Material and Methods

The subjects were 765 rheumatoid arthritic, 715 tubercular and 1,050 psychiatric outpatients treated by special society-supported outpatient clinics throughout Finland. The samples were randomly selected from patients visiting the clinics. The overall return rate of the questionnaires was above 70 %. However, some of the outpatient clinics which originally agreed to cooperate did not do so. When the units not giving the questionnaires to their patients were excluded from the material, the return rate of the patient groups was above 90 %. The control group of 587 persons was matched with the patient groups as to age range, which was between 18 and 70 years, and living district. The control group did not

[1] This study was supported by Liikenneturva, and the Finnish Antituberculosis Association, Helsinki.

Table I. Driving licence ownership among the rheumatoid arthritic, tuberculous and psychiatric subjects

Driving licence ownership	Rheumatoid arthritic group % (n = 765)	Tuberculous group % (n = 715)	Psychiatric group % (n = 1,050)	Control group % (n = 587)
Yes	37	59	29	58
No	63	41	71	42
Total	100	100	100	100

include subjects from districts of residence where the open care units did not cooperate. Control subjects were selected by sampling from the censuses of the countries having the outpatient clinics. The ages of the subjects were within the legal age-range for possessing a driving licence in Finland.

Results

Characteristics of the Subjects (table I)

Even though the groups differed from each other in many respects as might be expected from the epidemiology of the diseases, subjects possessing a driving licence proved relatively similar. Within all groups the majority were men aged 30–49 years. Males represented 70 % of drivers in the control group, 89 % of drivers in the tubercular group, and approximately 80 % in both the psychiatric and rheumatoid arthritic groups.

The social status of the drivers was generally similar in the patient and control groups. A significant exception was that 33 % of the drivers in the psychiatric group were of the lowest social class whereas the corresponding figure in the reference group was 13 % ($p < 0.01$).

Driving Characteristics (table II, III)

Significantly fewer subjects in the rheumatoid arthritic and psychiatric groups possessed a driver's licence than in the control group. The tubercular group did not differ from the controls.

The typical number of driven kilometres in the rheumatoid arthritic and psychiatric groups ranged from 1 to 9,999 annual kilometres. This was less than the annual driving experience of the tubercular and control subjects which was about 20,000 km or more.

The patient groups reported significantly more often than the control group that the type of route mainly used was highway. The main reason for driving was

Table II. Driving characteristics of rheumatoid arthritic, tuberculous and psychiatric subjects

	Rheumatoid arthritic group, %	Tuberculous group, %	Psychiatric group, %	Control group, %
Annually driven kilometres				
0	9	4	12	4
11–9,999	38	25	40	34
10,000–19,999	27	22	25	26
20,000 +	26	49	23	36
Total	100 (n = 229)	100 (n = 357)	100 (n = 239)	100 (n = 326)
Type of route				
Street, private road	29	21	22	37
Highway	71	79	78	63
Total	100 (n = 219)	100 (n = 378)	100 (n = 227)	100 (n = 328)
Type of driving				
Recreational driving	62	51	68	55
Driving to work	45	56	43	52
Chauffeurs	61	18	11	35
	(n = 214)	(n = 358)	(n = 221)	(n = 332)

in many cases difficult to specify. Many subjects indicated two categories, because they were not sure which one was most appropriate and therefore the percentage figures total more than 100. The main differences between the control and patient groups were that the patient groups included a significantly small number of chauffeurs and the rheumatoid arthritic and psychiatric patients did a significantly large amount of recreational driving. This finding was statistically significant.

The difference in the number of traffic accidents during the 2 years preceding the study was not significant between tubercular and control groups. The rheumatoid arthritic subjects had been involved in traffic accidents less often and the psychiatric group more often than the controls.

The correlation between annual driven kilometres and traffic accident involvement was positive. The rheumatoid arthritic group had the lowest accident involvement figures and the psychiatric subjects had the highest figures. It is to be specially noticed that the relative proportion of low-experienced psychiatric drivers is high among the accident-involved drivers. It is also worth mentioning

Table III. Traffic accident involvement of rheumatoid arthritic, tuberculous and psychiatric subjects according to the annual driven kilometres

Involvement in traffic accidents	Rheumatoid arthritic group			Tuberculous group		
	1–9 km (n = 88)	10–19 km (n = 58)	20 + km (n = 57)	1–9 km (n = 89)	10–19 km (n = 77)	20 + km (n = 167)
Yes	19	23	28	29	25	38
No	81	77	72	71	75	62
Total	100	100	100	100	100	100

Involvement in traffic accidents	Psychiatric group			Control group		
	1–9 km (n = 91)	10–19 km (n = 60)	20 + km (n = 50)	1–9 km (n = 112)	10–19 km (n = 85)	20 + km (n = 114)
Yes	35	44	49	24	35	48
No	65	56	51	76	65	52
Total	100	100	100	100	100	100

that although the tubercular subjects drove as much as the control group, the accident involvement rate of the tubercular drivers was lower than that of the controls.

Characteristics of Drug Use

The use of prescribed drugs was high in all patient groups, being on the level of about 90 % among rheumatoid arthritic and psychiatric patients. The corresponding figure for the tubercular group was 82 %. All these figures are significantly higher than the 41 % of the control group using some kind of prescribed drugs.

All patient groups had used drugs regularly every day during the previous 3 years, usually one or two drugs at a time. Specially the duration of drug treatment in the tubercular group was particularly long, generally longer than in the other patient groups.

Characteristics of Alcohol Use

The patient groups generally did not differ from the control group in their choice of alcohol, mixed beverages being generally preferred. However, the rheumatoid arthritic group showed a relatively greater preference for hard liquors and the psychiatric group for light alcohol, mostly beer or wine. In the patient groups there were relatively more nondrinkers than in the control group.

The rheumatoid arthritic subjects used alcohol less often and also drank less alcohol per drinking session than the controls. A similar tendency was evident when comparing the tubercular group with the control group. The psychiatric subjects drank as much and as often (once a week) as the controls.

Driving Characteristics and Drug Use (table IV)

In all patient groups the use of drugs was in inverse proportion to the number of annual kilometres driven. This tendency was clearest in rheumatoid arthritis and psychiatric groups. In the psychiatric group only 16 % of heavy drug users drove 20,000 km or more per year. The corresponding figure in the tubercular group was 41 % which is not significantly higher than in the control group.

In the rheumatoid arthritic and psychiatric groups the increasing drug use and the great amount of recreational driving seemed to be related tendencies. Among tubercular and psychiatric patients the increasing drug use and driving to work correlated inversely with each other. Drug use and accident involvement had no correlation in any group. The subjects most often involved in accidents used one or two drugs. This may be due to the fact that their frequency was highest in every patient group.

Driving Characteristics and Alcohol Use (table V)

In all patient groups heavy alcohol intake during one drinking session seemed to correlate with driving experience. As the alcohol intake per drinking session increased, the number of kilometers driven annually increased too. Conversely, the nondrinkers in each patient group drove little, only 1–9,999 km/year.

The tubercular, psychiatric and control nondrinker subjects drove mostly for recreation. Particularly in the psychiatric group the increase of alcohol intake during one drinking session related to the decrease in the amount of recreational driving. As a rule, drinking had no great effect on recreational driving.

Driving to work and the amounts of alcohol drunk during one drinking session seemed to be in direct proportion to each other. The increase in alcohol consumption per drinking session was also related to an increased tendency to drive to work. This correlation was strongest in the tubercular group. In the control group, driving to work was distributed equally to nondrinkers and drinkers.

Excluding the psychiatric nondrinkers, the nondrinkers were usually less frequently involved in traffic accidents than others. The frequency of accident involvement was found to increase in direct proportion to the amount of alcohol consumed per session. This may, however, be due to the fact mentioned above,

Table IV. Driving characteristics of the rheumatoid arthritic, tuberculous and psychiatric subjects according to the quantitative use of drugs

	Amount of drugs in the different groups											
	rheumatoid arthritic group			tuberculous group			psychiatric group			control group		
	0 %	1–2 %	3 + %	0 %	1–2 %	3 + %	0 %	1–2 %	3 + %	0 %	1–2 %	3 + %
Annually driven kilometres												
1–9,999	6 (n = 1)	40	55	23	28	37	52	42	60	31	29	67 (n = 10)
10,000–19,999	41 (n = 7)	30	24	23	23	22	11	32	24	31	19	–
20,000 +	53 (n = 9)	30	21	54	49	41	37	26	16	38	52	33 (n = 5)
Total	100 (n = 17)	100 (n = 111)	100 (n = 66)	100 (n = 48)	100 (n = 160)	100 (n = 74)	100 (n = 19)	100 (n = 111)	100 (n = 83)	100 (n = 162)	100 (n = 139)	100 (n = 15)
Type of route												
Street, private road	27	27	36	25	20	23	27	21	25	37	42	47 (n = 7)
Highway	73	73	64	75	80	73	73	79	75	63	58	53 (n = 8)
Total	100 (n = 18)	100 (n = 116)	100 (n = 72)	100 (n = 49)	100 (n = 168)	100 (n = 79)	100 (n = 22)	100 (n = 117)	100 (n = 89)	100 (n = 164)	100 (n = 99)	100 (n = 15)

Table IV (continued)

	Amount of drugs in the different groups											
	rheumatoid arthritic group			tuberculous group			psychiatric group			control group		
	0 %	1–2 %	3 + %	0 %	1–2 %	3 + %	0 %	1–2 %	3 + %	0 %	1–2 %	3 + %
Type of driving												
Recreational driving	55 (n = 10)	56	70	62	52	53	67	64	79	48	61	56 (n = 7)
Driving to work	39 (n = 7)	54	17	62	58	48	48	48	30	55	44	69 (n = 9)
Chauffeurs	32 (n = 6)	16	18	21	14	19	5	13	9	15	13	30
	(n = 18)	(n = 112)	(n = 72)	(n = 48)	(n = 169)	(n = 75)	(n = 21)	(n = 111)	(n = 80)	(n = 165)	(n = 101)	(n = 16)
Involvement in traffic accidents												
Yes	49 (n = 8)	20	22	26	32	36	30	46	35	27	42	27 (n = 4)
No	51 (n = 9)	80	78	74	68	64	70	54	65	73	58	73 (n = 11)
Total	100 (n = 17)	100 (n = 110)	100 (n = 71)	100 (n = 49)	100 (n = 167)	100 (n = 77)	100 (n = 23)	100 (n = 119)	100 (n = 94)	100 (n = 163)	100 (n = 98)	100 (n = 15)

Table V. Driving characteristics of the rheumatoid arthritic, tuberculous and psychiatric subjects according to the quantitative use of alcohol

	Amount of alcohol in the different groups											
	rheumatoid arthritic group			tuberculous group			psychiatric group			control group		
	0 g%	1–49 g%	50+ g%	0 g%	1–49 g%	50+ g%	0 g%	1–49 g%	50+ g%	0 g%	1–49 g%	50+ g%
Annually driven kilometres												
1–9,999	60	46	23	32	44	18	53	58	40	48	36	32
10,000–19 999	24	27	37	28	18	23	29	29	25	26	31	26
20 000 +	16	27	40	40	38	59	18	13	35	26	33	42
Total	100	100	100	100	100	100	100	100	100	100	100	100
	(n = 67)	(n = 67)	(n = 73)	(n = 98)	(n = 91)	(n = 152)	(n = 55)	(n = 48)	(n = 116)	(n = 48)	(n = 87)	(n = 180)
Type of route												
Street, private road	28	31	31	31	19	33	14	23	25	29	38	40
Highway	72	69	69	69	81	67	86	77	75	71	62	60
Total	100	100	100	100	100	100	100	100	100	100	100	100
	(n = 71)	(n = 70)	(n = 72)	(n = 75)	(n = 103)	(n = 95)	(n = 56)	(n = 52)	(n = 129)	(n = 49)	(n = 90)	(n = 184)
Type of driving												
Recreational driving	60	67	58	57	42	51	80	73	61	68	46	52
Driving to work	46	36	52	51	60	72	35	37	48	46	57	51
Chauffeurs	18	16	17	16	19	18	10	4	14	9	17	87
	(n = 67)	(n = 69)	(n = 73)	(n = 103)	(n = 85)	(n = 169)	(n = 49)	(n = 51)	(n = 122)	(n = 12)	(n = 91)	(n = 187)
Involvement in traffic accidents												
Yes	12	26	26	26	31	38	40	26	49	19	33	41
No	88	74	74	74	69	62	60	74	51	81	67	59
Total	100	100	100	100	100	100	100	100	100	100	100	100
	(n = 76)	(n = 72)	(n = 73)	(n = 107)	(n = 84)	(n = 156)	(n = 58)	(n = 62)	(n = 124)	(n = 47)	(n = 90)	(n = 185)

that heavy drinkers on the average drive more than the others. In the psychiatric group accident involvement did not seem to depend on alcohol intake. In this group the accident statistics were high both in drinkers and nondrinkers.

To summarize the main results, the rheumatoid arthritic and psychiatric subjects tended to drive less than tubercular and control subjects. The rheumatoid arthritic patients were involved in traffic accidents less often, and the psychiatric patients more often than other groups.

The tubercular and control subjects tended to have a lot of driving experience but tubercular patients in particular, had a low accident rate. The increase of drug use correlated inversely with annual kilometres driven. Alcohol intake per session increased in direct proportion both to annual kilometres driven and to accident involvement.

Discussion

All questionnaire studies about drinking and driving have had severe limitations (4). There have been no studies to answer the question: How much of the impairment of skills observed among the patient groups was due to disease, how much to their treatment and how much to the interaction between these factors? The present study was undertaken in order to elucidate these points by means of comparing the nondrug patient groups with controls. However, it should be kept in mind that a heavy medication generally reflected a severe disease, and the role of the illness itself in impairing driving skills remains somewhat obscure. The reliability of the present results is believed to be high, due to the high percentage of the subjects who returned the questionnaire and the strict confidentiality of the data.

The relatively low number of driver's licences held by the rheumatoid and psychiatric patient groups might result from three factors. (1) The disease affected the persons severely enough to make them reluctant to drive. (2) Income was so low due to the disease that the person could not afford to drive. (3) The health care personnel had recommended discontinuing driving due to the disease or police had confiscated the licence for the same reason.

We consider that all these possibilities contributed to the driving in the rheumatoid arthritic group, consisting mainly of elderly women. Even those in the rheumatoid arthritic group who held driving licences drove relatively less than control subjects. This probably reflected an impairment either due to the disease or its treatment. This low number of kilometres driven annually meant that the rheumatoid arthritic and psychiatric outpatients were less experienced in the present traffic conditions than the tubercular and control subjects. This may account for their relatively high accident-involvement rate.

The use of alcohol as indicated by the amount of pure alcohol ingested per drinking session did not differ greatly from one group to another. In all groups consumption of alcohol increased in direct proportion to amount of driving undertaken, except among psychiatric patients where the effect of heavy alcohol use on driving seemed to be bimodal. The psychiatric nondrinkers were involved in accidents more often than the corresponding control subjects whereas among the control group, rheumatoid arthritic and tubercular patients the heavy use of alcohol showed a more positive correlation with accident involvement, than the nonuse of it. This increased accident risk was not real but correlated with the increased mileage, being thus mainly due to increased exposure to traffic.

The relatively low rate of accident involvement among the heavy drinkers of the rheumatoid arthritic group could be explained, for example, as follows: (1) rheumatoid arthritic patients often tend to be depressive (5), so, the effect of alcohol on them may differ from that on other subjects; (2) some of the drugs used for the treatment of rheumatoid arthritis aggravated the effects of alcohol as may happen in the extreme conditions of a laboratory experiment (3).

An interesting finding was that 41 % of control group used some kind of medication. As to the correlation between drug use and accident rates by far the most important finding was that among those individuals in patient groups who did not use drugs, the accident rate was not higher than that among the corresponding control subjects. Furthermore, in the control and psychiatric groups the subjects taking no drugs were less involved in accidents that those having one or two drugs. Thus, drugs contribute to a significant degree to traffic accidents, and this is even true for drugs other than psychiatric ones. The fact that this was not the case with respect to rheumatoid arthritic patients could result from the treatment (frequently aspirin) which, in the laboratory, does not impair psychomotor skills related to driving (3).

Isoniazid, the most common antitubercular agent in Finland, may impair driving skills in a simulator study (2). The relevance of this finding to field conditions did not become apparent in the present study, and care should be taken when relating laboratory data to practice.

With respect to the combined use of drugs and alcohol, about half of the heavy drinkers, both control subjects and psychiatric outpatients using one or two drugs, had been involved in accidents during the 2 years previous to the study. The high figure points out the conclusion that although heavy drinking correlated with greater exposure to traffic in the present data, in combination with drugs there was an extra accident risk factor.

Summary

A questionnaire was administered to 765 rheumatoid arthritic, 715 tubercular and 1,050 psychiatric outpatients and to 587 control subjects concerning their use of drugs and alcohol, driving habits and traffic accident involvement. The driving populations of all groups were matched as to their age and area of residence.

The results of the study show a relatively very small number of driving licence holders among rheumatoid and psychiatric patients whereas the tubercular group includes as many driving licences as the control group. Also, the annual driving experience of the rheumatoid arthritic and psychiatric patients was low. The main reason for driving was recreation, and the driving mainly took place on highways. Excluding the psychiatric patients, the patient groups had been involved less often in traffic accidents than the controls.

The use of drugs correlated inversely with driving: the more you used drugs the less you drove. Also the consumption of alcohol increased in direct proportion to the number of kilometres driven: the greater the alcohol consumption per drinking session, the more you drove annually and were involved in accidents.

References

1 *Bo, O.; Haffner, J.F.W.; Langard, O.; Trumpy, J.H.; Bredesen, J.E., and Lunde, P.K.M.:* Ethanol and diazepam as causative agents in road traffic accidents. Proc. 6th Int. Conf. on Alcohol, Drugs and Driving, Toronto 1974.

2 *Linnoila, M. and Mattila, M.J.:* Effects of isoniazid on psychomotor skills related to driving. Clin. Pharmac. *13:* 343–350 (1973).

3 *Linnoila, M.; Seppälä, T., and Mattila, M.J.:* Acute effect of antipyretic analgesics alone or in combination with alcohol, on human psychomotor skills related to driving. Br. J. clin. Pharmac. *1:* 472–484 (1974).

4 *Nichols, J.L.:* Drug use and highway safety. Review of the literature. DOT-HS-371-3-786. US Dept. of Transportation (1971).

5 *Rauhala, U.:* Suomalaisen yhteiskunnan sosiaalinen kerrostuneisuus, WSOY, Helsinki 1966.

Dr. *M. Mäki,* Department of Research Liikenneturva, Iso-Roobertink. 20, *SF–00120 Helsinki 12* (Finland)

Mod. Probl. Pharmacopsych., vol. 11, pp. 22–41 (Karger, Basel 1976)

Alcohol and Highway Crashes

Closing the Gap between Epidemiology and Experimentation[1]

M. W. Perrine

Project ABETS, Psychology Department, University of Vermont, Burlington, Vt.

One night, as a young woman was crossing the main road through her village, a car suddenly appeared. She quickly turned her head and looked fearfully at the rapidly approaching headlights. Then, the tires screeched, but the car skidded out of control into the woman. The driver staggered out of the car and stared helplessly at the motionless woman in the road. He kept shaking his head in disbelief and repeating, 'I just couldn't stop in time; I just couldn't stop fast enough!'

This man had been drinking and was legally impaired according to the law. He was arrested for driving while intoxicated (DWI) by the police officer who arrived shortly after the crash.

This fictitious anecdote is an analog of the type of incident that has been repeated countless times throughout modern nations since the early years of the automobile age. Actual incidents of this type have provided the tragic data for the epidemiologic study of alcohol and highway crashes. However, for the purposes of the present paper, the major influence of such incidents over the years has been the resulting emphasis on the association between alcohol and 'not being able to stop the car in time'. This observed association led to the inference that alcohol simply slows the reaction itself, that is, the motor response of applying the brakes. Although this oversimplification still persists

[1] All Project ABETS research by the writer and his associates reported herein was supported by the National Highway Safety Administration of the US Department of Transportation (under contracts FH-11-6606, FH-11-6899, FH-11-7469, and DOT-HS-364-3-757) and/or by the National Institute on Alcohol Abuse and Alcoholism of the US Department of Health, Education, and Welfare (under grants MH-17583, and AA00246-05). The opinions, findings, and conclusions expressed in this publication are those of the writer and not necessarily those of the National Highway Traffic Safety Administration or the National Institute on Alcohol Abuse and Alcoholism.

widely, it has nevertheless stimulated much experimentation over the years, with primary emphasis on alcohol influences upon reaction time and gross psychomotor responses.

Although the statement 'I couldn't stop in time' is correct in a literal sense for many alcohol-involved crashes, it is only a *description* of what was observed – and is not an *explanation* of what happened. Thus, even though the statement is true as a description or an observation, it does not provide any understanding of the basic problem: exactly *what* aspect of driving behavior is being influenced by alcohol. In other words, alcohol could be impairing *any* stage or combination of stages in the behavioral sequence which begins with the sensory input and ends with the response output. This sequence can be conceptualized in terms of information processing, and as such, is discussed below.

The Gap between Epidemiology and Experimentation

In attempting to understand the basic problem concerning the exact contribution of alcohol to highway crashes, there is still a large gap between description and explanation which at this point in time, can only be bridged provisionally through the use of inferences. Two widely separated research approaches have been used to date as a basis for inferring the contribution of alcohol to highway crashes: *epidemiologic* and *experimental.*

First, high blood alcohol concentrations (BAC) have been thoroughly implicated in serious and fatal injury highway crashes by post hoc epidemiologic studies. Most evidence for inferring this alcohol contribution to highway crashes has been obtained by examining the distribution of BAC among drivers involved in actual crashes (both fatal and nonfatal) and/or among drivers using the highways but not involved in crashes at the time (on the basis of roadside surveys). A number of such case/control studies have demonstrated that alcohol is overrepresented among deceased drivers relative to drivers in the population-at-risk using the highways at corresponding times and places (6, 17, 18, 25, 30, 35, 36, 39, 40).

The second approach consists of controlled experiments conducted on isolated variables; and it has been the practice to infer alcohol impairment of real-world driving performance from the mosaic of fragmented bits of behavior examined separately in the laboratory and in part-task driving simulators (e.g., reaction time, tracking, etc.). In addition to the problems involved in reasoning from isolated parts to a complex whole, several aspects of human behavior may set further limitations on the extent to which we can extrapolate from controlled experiments. For example, one limitation stems from the 'grandstand effect' or Hawthorne effect in which the subject who knows that he is under observation may exert himself and compensate for the effects of alcohol, especially at lower BACs. One serious consequence of this effect is that any attempt to establish criteria for alcohol impairment based upon this aroused, grandstand performance could err seriously in the direction of setting the standards for impairment far too high for the extent actually involved in unobserved natural-state driving on the road, during which the depressant effects of alcohol would assumedly be more dominant. One possible solution for this problem is to develop valid, unobtrusive measures of alcohol impairment in real-world driving situations. One such attempt is discussed below.

In the meantime, a pragmatically acceptable compromise consists of inferring real-world alcohol impairment of those behavioral variables which show *consistent* degradation in controlled experiments. In these terms, the basic problem concerns those aspects of actual driving performance which are *differentially* impaired by alcohol; more specifically, those alcohol-induced changes in driving behavior which would serve to differentiate motorists with high BACs from motorists with zero or low BACs. To date, the most dependable information bearing upon this question has been obtained from systematic experiments conducted in the laboratory or in cars driven either on closed-course driving ranges or on actual public roads. The results of these experiments have been reviewed in a number of recent literature surveys (1, 2, 9, 11, 13, 19, 22–24, 26–29, 38).

On the basis of these experiments, the following aspects of driving-related behavior appear to be consistently and unequivocally impaired by alcohol: *divided attention, compensatory tracking,* and *choice reaction time.* Alcohol impairment of these aspects is reflected in the following driving variables: *steering reversal, lateral position, speed changes, speeding,* and *braking.* All these variables have been systematically studied in closed-course driving experiments and are influenced by alcohol (11, 13, 33).

Thus, if the assumption is correct that alcohol actually does degrade a driver's capabilities and performance, then alcohol-induced changes in on-road driving behavior should be manifest in some fashion; accordingly, these changes should be amenable to systematic observation and recording. Nevertheless, few empirical data are available to provide *direct* support for this assumption, i.e., no controlled study is known to have been conducted previously to obtain systematic but unobtrusive data on the *actual* influences of alcohol upon real-world driving behavior in its natural environment. Recently, however, we have completed a study in Vermont which was designed to begin filling this gap between epidemiology and experimentation – or, in one sense, the gap between description and explanation. Unfortunately, time limitations permit consideration of only one of the performance measures obtained in this study, and I have selected the one variable which lends itself most readily to comparison across the full spectrum of alcohol experimentation, namely, braking and stopping the car 'in time'.

Reaction Time and Brake Use

In order to begin closing the gap between epidemiology and experimentation, it is instructive to examine alcohol influences upon a single variable across as many observation points as possible, ranging from simple laboratory experiments, to part-task driving simulator experiments, to closed-course driving experiments using instrumented cars, to on-road investigations using experimental subjects, to unobtrusive measures of actual on-road driving performance of the motoring public, and to after-the-fact epidemiologic and on-the-scene accident investigation studies. *Reaction time* is one of the most frequently used measures in the behavioral experiments. *Braking-and-stopping* is one of the most frequently used maneuvers in attempting to avoid or minimize a crash. Thus, it seems most appropriate to select these two variables and their interrelations as the focus for the present paper.

Braking is the final 'element' or stage in the information-processing sequence involved in bringing a car to a stop. However, as noted above, it should not necessarily be assumed that alcohol simply slows the reaction itself, that is, the motor response of applying the brakes. Rather, alcohol could be affecting any or all stages of the information-processing sequence. Since both braking and reaction time are more readily observable and measurable than aspects of the other preceding stages in this sequence, relatively more data are available for the purposes of comparison across the spectrum from epidemiology to experimentation. Accordingly, in the remainder of the present paper, let us examine alcohol influences (1) upon reaction time as investigated in laboratory experiments (including part-task simulator experiments), (2) upon reaction time as investigated in instrumented car experiments, (3) upon braking performance in instrumented car experiments, and (4) upon braking performance in our recent field experiment. Unfortunately, space limitations permit consideration of only one or two examples in each of these four categories.

Reaction Time in Laboratory Experiments

Reaction time is generally increased by alcohol; however, a number of qualifications are necessary. The two most important considerations are the relative amount of alcohol involved and the type of reaction time task involved, i.e., the most basic dichotomy being whether the task involves simple as opposed to choice reaction time. These considerations have been specifically addressed in several literature reviews (5, 22, 23, 38).

For our present purposes, it is sufficient to note that despite inconsistencies and considerable disagreement among the many published studies, BACs above 80 mg% (= 80 mg/ml or 0.8 ‰) must be achieved before *consistent* increases in reaction time are obtained. Furthermore, *simple reaction times* have been relatively resistant to alcohol impairment and have rarely been increased by more than 10% – if at all. Changes of such relatively small magnitude have been casually dismissed by some investigators as being negligible and as having little practical significance. By contrast, however, studies involving *choice reaction times* have generally yielded greater alcohol impairment and at lower BACs, although the range of impairment is relatively large.

Several recent attempts to reduce this disagreement have been based upon an information-processing approach (10, 14, 22, 23, 37). In almost all these experiments, response latency or reaction time was taken as the primary indicator of information processing, and various strategies or models were then invoked in the attempts to attribute reaction time changes to alcohol influences upon implicated aspects of the information-processing sequence. However, few attempts have been made to apply any systematic model in order to specify more precisely the locus of alcohol effects within the information-processing sequence.

In one example, *Huntley* (14) used *Smith*'s (34) four-stage model of information processing in attempting to determine the differential alcohol sensitivity of the two traditional dichotomous aspects of processing: *input* (stage 1: stimulus registration and detection; and stage 2: stimulus recognition), and *output* (stage 3: response selection; and stage 4: peripheral response implementation). To enable differentiation between input and output, he separated stimulus effects from response effects by manipulating stimulus-response familiarity and measuring reaction time. He found that when the associated novel responses were required, reaction time was significantly lengthened by alcohol and the magnitude of this effect was significantly enhanced by increasing stimulus-response uncertainty. *Huntley* (14) concluded that 'the locus of the alcohol effects in the information-processing sequence appears to be in the stimulus-response translation process, i.e., in response encoding rather than with stimulus recognition *per se* or response execution'.

In a recent review of the alcohol and information processing experiments, *Perrine* (27) concluded that at medium BACs, information processing is relatively resistant to alcohol impairment when both stimulus and response conditions are completely congruent or at least are highly familiar; but, by contrast, alcohol impairment of information processing increases as the required complexity of processing increases. He also concluded that reaction times manifest relatively little alcohol impairment in tasks involving high stimulus-response association strength and requiring responses which are straightforward, compatible, and highly familiar (27, p. 23).

Regarding alcohol effects upon behaviors measured in *driving simulators, Heimstra and Struckman* (9) concluded that it was primarily associated with variability of performance, both within and between studies. Nevertheless, the authors concluded that the performance changes on tasks involving complex reaction time appear to have been more consistent than the changes in motor performance. In fact, this relative consistency – along with the indication of the importance of complex, higher level tasks in some of the studies – led to the conclusion that 'perhaps the most important factor determining the impairment of the driving task is the effect of alcohol on the higher mental processes' (9, p. 27).

In this regard, it is especially noteworthy that the authors found a majority of the experiments reviewed had only investigated behaviors relating to the 'physical response output task'. This is the fourth and final stage in the information-processing model mentioned above (14). For at least two reasons, this finding more than any other single factor probably accounts for the general absence of any consistent alcohol impairment on the behaviors studied with driving simulators. First, according to the review by *Levine et al.* (19), psychomotor tasks appear to suffer the least alcohol impairment by comparison with cognitive tasks and perceptual-sensory tasks which appeared to be most impaired. Secondly, according to *Huntley*'s (14) investigation, alcohol effects upon the information-processing sequence do not appear to be localized at the response execution stage (or the stimulus recognition stage), but rather at the response encoding stage. Thus, in conjunction with the interpretations offered above, these two sources of evidence can account for the results upon which *Heimstra and Struckman* (9) based their conclusions. Furthermore, on the basis of the simulator studies published since their 1972 review, *Perrine* (27) concluded that there was no reason for modifying these conclusions.

Reaction Time in Instrumented Car Experiments

Despite its apparent importance and relative ease of measurement, reaction time has been used in only a few controlled experiments conducted with instrumented cars and alcohol. However, it should also be emphasized that the total number of such studies is relatively small, probably no more than two dozen. In the definitive review of alcohol influences upon closed-course driving performance using instrumented cars, *Huntley* (11, 13) reported on three studies which used *brake reaction time* as a dependent variable, as opposed to other measures of braking performance such as stopping accuracy, stopping distance, brake pressure, etc. (variables which are discussed in the next subsection).

A study by *Chastain* (7) was one of the first in which alcohol influences were investigated at speeds and on tasks which were relevant for on-road conditions. While driving at 20 mph (32 km) through certain portions of a closed-course, an auditory signal was presented and the car was to be stopped as quickly as possible. Brake reaction times were increased 17 % in the alcohol condition relative to a no-alcohol condition the previous day (BACs ranged from 100 to 120 mg% across the six subjects).

The only other controlled experiments known to be relevant are three in a series of eight investigations we have conducted at the University of Vermont in recent years using our first instrumented car (33). In one study (discussed in the next section), we wished to examine the influences of alcohol upon *functional* braking (and steering) responses as well as upon brake-pressure modulation during a driving task which simulated a passing maneuver (16, 33, ch. 8). However, the response required in the other two studies was a light tap of the brake pedal which was nonfunctional in the sense that the pedal was literally only tapped and a conscious attempt was made to avoid changing the car speed as a result. These two driving experiments were designed primarily to pursue our continuing interest in alcohol influences upon the allocation of attention and effort in the foveal-extrafoveal paradigm. Two different types of attentional subsidiary tasks were used, involving signal detection in the first experiment and letter recognition in the second. Due to time limitations, however, the second study cannot be reviewed here.

The first study (15, 33, ch. 6) was conducted to investigate alcohol influences upon the ability to divide attention between a speed-maintenance task and detecting foveal and extrafoveal light signals while driving. Of particular interest was the extent to which alcohol would increase *simple* reaction time in a relatively realistic driving situation. The influences of the two additional, concurrent tasks upon control-use performance and heart rate were also examined. In order to reduce the effects of between-subject variability and of arousal (which tends to offset alcohol impairment), a very small number of subjects was tested on many occasions using a rather boring course and driving task, as well as very infrequent stimulus presentation in the light-signal detection task.

Two male volunteers served as paid subjects on nine nights in a 2 × 2 × 2 factorial design involving target BACs of 0 and 100 mg%, two light-signal locations (0 and 60°), and two driving speeds (15 and 25 mph) (24 and 40 km). A placebo beverage was consumed on four nights, alternating with consumption of the alcohol beverage on the remaining four nights.

On each of the eight nights, each subject drove four practice and four test laps over a 2.7-mile (4.4 km) circuit of deserted two-lane roads at the required speed of either 15 or 24 mph (24 or 40 km). Prior to each lap, the subject was instructed that the primary task was to keep the car at the designated speed for that lap, and secondly that he should respond to the light signals as fast as possible by removing his right foot from the accelerator and depressing the brake pedal slightly. Three dim light signals were presented from the center of the visual field or from 60° to the left and right of center on the horizontal meridian.

Baseline measures of heart rate and reaction time for signal detection were recorded first while sitting in the parked instrumented car. After drinking an alcohol or a placebo beverage, these measures were again obtained in the stationary mode, as well as while each subject drove the course four times at either 15 or 25 mph (24 or 40 km). Control-use measures were also obtained during

Table I. Means of simple reaction times (msec) to foveal and extrafoveal signals according to alcohol and placebo conditions for driving and parked modes

Condition	Foveal	Extrafoveal
Driving		
Alcohol	909	1,152
Placebo	744	868
Parked		
Alcohol	678	791
Placebo	572	660

the driving trials, after which heart rate and reaction time data were again recorded in the stationary mode.

Reaction times were significantly increased by alcohol, light-signal location, and driving (table I). In comparative terms, reaction times to the target lights were: 17 % slower to extrafoveal than to foveal signals while driving without alcohol; 22 and 33 % slower to foveal and extrafoveal signals, respectively, while driving after alcohol; approximately 31 % slower while driving with no alcohol than while parked; and 46 % slower to extrafoveal after alcohol while driving than while parked. The slowest reaction times were produced in detecting the extrafoveal lights in the alcohol/driving condition, whereas the fastest were with the foveal light in the placebo/parked condition.

The relatively great magnitude of the alcohol effect upon reaction time may be at least partially attributable to the nature of the required response; that is, the comparatively gross movement of shifting a foot from the accelerator to the brake pedal involved in this study is admittedly different from the finger movement typically used as the required response in most laboratory studies. Regardless of this difference, however, the main point is that this foot response is the one actually used for braking in real-world driving. Therefore, these results are much more applicable to actual driving situations than results obtained in laboratory studies.

Furthermore, the real-world implications of these 22 and 33 % increases in reaction time under alcohol become especially compelling when even this conservative estimate (i.e., derived from well-practiced experimental subjects) is translated into proportional increases in stopping distance at the higher speeds typically encountered on public highways. For example, if a DWI motorist (100 mg%) were driving at 55 mph (89 km), it would take him at least 1.3 sec longer to stop in response to a *foveal* signal (e.g., something right in the middle of his lane) than if he were sober. In terms of distance, his car would have traveled 47 ft (18.7 m or almost 3 car lengths) farther during this time than if he had been sober. However, in response to an *extrafoveal* signal (e.g., a child or a car coming out from the side of the road), it would take him at least 2 sec longer to stop than if he were sober. During this time, his car would have traveled 70.5 ft (28 m or 4 car lengths) farther than if he had been sober.

It was concluded that these slower reaction times to the light signals while actually driving the car under alcohol reflect a reduction in available attention to allocate to competing concurrent tasks. These combined considerations provide important evidence for the specific alcohol impairment of driving.

In the second of these two related studies (33, ch. 7), analogous results were obtained. However, of particular relevance for present purposes was the significant difference in reaction time as a function of response mode. Alcohol increased reaction time for foot responses (brake, or accelerator) significantly more than for hand responses (steer right, or steer left).

In summary, it seems clear from these studies that brake reaction time is significantly impaired by alcohol (ranging around 100 mg%) in a variety of subsidiary tasks performed by subjects while driving instrumented cars.

Brake Use in Instrumented Car Experiments

Useful measures of braking performance other than reaction time include: stopping *accuracy*, stopping *distance*, brake *pressure*, brake-pressure *modulation*, and braking *smoothness*. One or more of these variables has been used in each of the three known studies in which alcohol influences upon braking performance has been investigated using instrumented cars. These three studies (16, 20, 32) have been extensively reported and compared by *Huntley* in his recent review (11, 13). Alcohol impairment of various aspects of braking performance was found in all three studies.

As one part of their well-designed closed-course experiment, *Lovibond and Bird* (20) examined alcohol influences upon braking performance by using accelerometers, as well as a measure of stopping distance. To obtain the latter measure, the observer riding in the car with the subject triggered a blank charge mounted under the bumper which fired yellow powder onto the roadway and simultaneously provided the driver with an auditory signal to apply the brakes. As soon as the driver depressed the brake pedal, it triggered a second blank which also fired a charge of the yellow powder onto the road surface. The distance between the two spots of yellow powder provided a measure of brake response time. Although relatively little information concerning the braking task and its results was provided in the original paper, *Huntley* (11, 13) has calculated that braking *distance* increased 35 % in the 100 mg% alcohol condition relative to the placebo condition.

Stopping *accuracy* was measured in our very first instrumented car experiment (32, 33, ch. 3), as well as in the *Lovibond and Bird* (20) study. In both studies, the subjects were required to halt the car as close as possible to specified stoplines and the measure consisted of the actual distance of the car from the line. *Lovibond and Bird* required the subjects to perform such a task while driving slowly in reverse, as well as while driving rapidly forward. The subjects were less accurate in stopping under alcohol, but no statistical tests were reported. In our experiment, the car speeds seldom exceeded 10 mph (16 km) at either of the two stoplines, but even so, the subjects stopped the car significantly less accurately in the 100 mg% condition than they did in the no-alcohol condition.

In our simulated passing experiment (16, 33, ch. 8), measures of *brake-pressure modulation* were obtained in addition to braking response time. It was expected that brake-pressure modulation would be influenced by alcohol because of its dependence upon muscle feedback – and consequently upon reaction time. Although the practical implications of increased reaction time during driving emergencies are obvious, the most important ramification of these effects may not be so apparent. For example, since the appropriate modulation of brake pressure in emergency situations is to some extent dependent upon reactions to proprioceptive feedback, it can be expected that alcohol will influence the *shape* of brake-pressure curves and thus braking efficiency, as well as cause the expected increase in time between signal detection and beginning of brake response. The shape of the brake-pressure curve and the braking efficiency would in turn be reflected in the smoothness and degree of control with which the car was slowed or brought to a stop. Thus, in the anecdote related at the beginning of the present paper, the drunken driver did not brake very efficiently, as indicated by the car skidding out of control. These aspects of braking performance obtained in controlled experiments are especially interesting in view of our field study of on-road driving behavior and the braking data we obtained by measuring the deceleration rate of each motorist as he was being signalled to stop by a state police officer (discussed in the following subsection).

In the simulated passing experiments, the task involved emergency control-responses in an aborted passing maneuver, as well as the more relaxed control-responses used in the typical return-to-lane maneuver. While driving slowly (10 mph) (16 km) in the right lane of a stylized two-lane road, each subject was required to pull out into the left lane to pass a barricade simulating the rear of a truck, accelerate rapidly past it and through a pylon-defined 'crossroad', and then return to the right lane upon receiving a light signal flashed in the rearview mirror. However, on 50 % of the trials, an abort signal was presented as the 'crossroad' was approached. This light signal appeared on top of either the front left or right fender and indicated that the car should be swerved respectively to the left or right side of the road and stopped quickly.

Sixteen male volunteers served as paid subjects in a 2 X 2 factorial design involving two target BACs of 0 and 100 mg% and two passing conditions (aborted or completed. Each subject drove the 900-ft (274-m) course 30 times on each

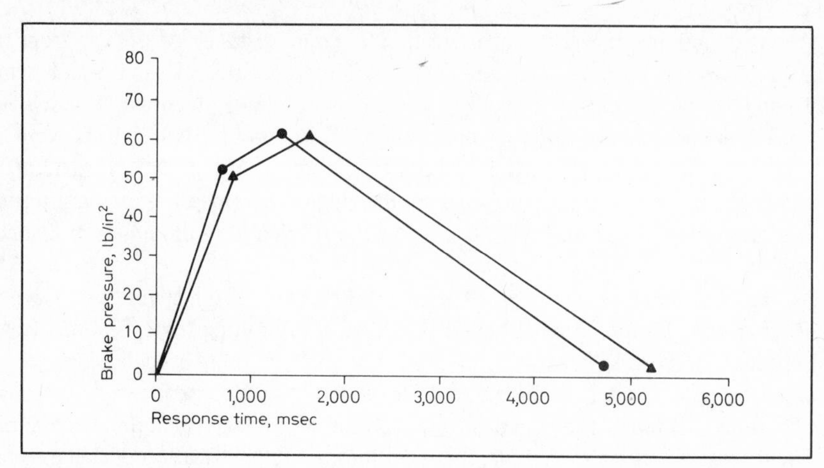

Fig. 1. Brake pressure (lb/in^2) as a function of reaction time (msec) for the abort trials. ● = Alcohol; ▲ = placebo.

of three separate nights, the first of which was used for practice. Each nightly session of 30 trials was evenly divided between abort and passing trials which were presented in random order.

Analysis of variance revealed that alcohol significantly increased steering and braking *response times* in both types of trials. Although statistically significant, the mean increase in response time under alcohol was only 65 msec or 6 %.

The *brake-pressure* data were obtained in analog form, examined visually, and then digitized in a manner intended to discriminate between alcohol- and placebo-associated curves, as well as to describe the essence of the curves. The values representing five separate measures were then summed across trials and analysis of variance conducted with the obtained means.

Schematic representations of the brake-pressure functions derived for the abort and passing trials are presented in figures 1 and 2, respectively. The three data points defining each function in these figures represent the mean time and amplitude of the initial pressure maximum, mean time and amplitude of maximum pressure produced, and the mean duration of the brake depression response. Each data point represents the mean of 192 responses.

When the two figures are compared, it can be seen that passing trials produced greater pressure amplitudes and longer brake depression durations than abort trials. However, only the amplitude differences were statistically significant. The amplitude differences were most likely the result of the higher driving speeds attained during the passing trials.

Regarding alcohol influences upon braking performance, both figures indicate that maximum amplitudes were obtained earlier and brake depressions were

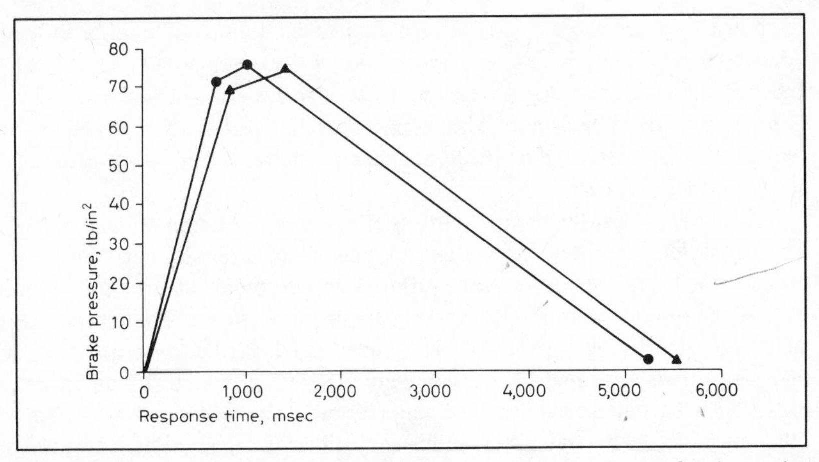

Fig. 2. Brake pressure (lb/in²) as a function of reaction time (msec) for the passing trials. ● = Alcohol; ▲ = placebo.

shorter on alcohol trials than on placebo trials. However, only the time from onset of braking to the time at which maximum brake pressure was reached was statistically significant. The mean decrease was approximately 25 %. Thus, under alcohol, rise times of the brake-pressure functions were shorter. The finding that alcohol decreased the rise times of the brake-pressure functions might appear to indicate an increase in braking efficiency and thus seem inconsistent with the reported effects of alcohol upon response times. However, since subjects were instructed to stop smoothly in all circumstances, the faster rise times actually reflect impairment of performance, being an indication of more abrupt braking.

In summary, braking and steering response times were increased by alcohol. Furthermore, the *shape* of the brake-pressure curves changed with alcohol, notably in that significantly faster rise times were recorded from the onset of brake pressure to the point in time at which maximum pressure was produced. The duration of braking was also shorter under alcohol. It was concluded that the more sharply peaked curves in the alcohol/passing condition reflected an abruptness and reduction in stopping smoothness and in the sensitivity with which the brakes were applied.

Braking Performance on Public Highways

Evidence concerning alcohol impairment of actual driving performance has been obtained from three types of situations: (1) more or less contrived driving on a closed course using experimental subjects; (2) controlled studies of on-road driving using experimental subjects, or (3) natural on-road driving. The closed driving course situation has been discussed in the two previous subsections.

Regarding the second type, several controlled studies of alcohol influences upon on-road driving have been reported (3, 4, 21), but in all cases the drivers knew that they were being used as subjects in an experiment under the observation of a research assistant. The alcohol-impaired variables reported in these studies included steering reversals, visual scan pattern, and spare capacity for divided attention tasks.

Until very recently, almost all evidence of alcohol impairment of natural on-road driving resulted from direct visual observation by law enforcement officers. Such evidence necessarily suffered an enormous intrinsic source of bias in that the officers only stopped a very small proportion of all drivers observed and, furthermore, measured BACs in an even much smaller proportion. Accordingly, motorists with high BACs who were manifesting erratic driving maneuvers had a much higher probability of being stopped and questioned by police than other motorists who did not manifest any deviant driving, but who may have had equally high or higher BACs.

Our very recent field study (8) represents an innovative exception to the traditional procedures, in that unobtrusive electronic measures of nocturnal driving performance were obtained for all motorists passing selected sites on public highways. All these motorists were subsequently stopped by a state police officer. Then, they were approached by members of the research team in order to obtain BAC measures, as well as interview data concerning biographical variables, drinking patterns, and driving records.

Basically, we were interested in determining whether the driving behavior of motorists with high BACs was discriminably different from that of motorists with low or zero BACs. We also wanted to determine whether such discriminable differences – if we obtained them – are also related to the motorist's reported drinking patterns and driving history. Our specific task consisted of developing and then testing some unobtrusive means of detecting drivers with high BACs while they are actually using the highways. Ideally, the resulting techniques might eventually be useful for enforcement as a type of what might be called an 'alcohol radar'. Thus, this research project was aimed at the longer range goal of reducing alcohol-involved crashes on the highways by providing a means of improving early, on-road detection of high-risk, high BAC drivers.

Since you have just seen our film, entitled *Roadside Research*[2] documenting the instrumentation and field operations in this project, I will now limit myself to the minimum description necessary for an understanding of the one performance measure selected for emphasis in the present paper, namely, *braking performance.*

[2] A film was presented to the Symposium (*Perrine et al.:* Roadside Research). Arrangements for showing the film can be made by writing to the author.

In an attempt to meet the objectives of the NHTSA contract, our original research plan was designed in terms of the following, very specific requirement in the work statement: this project will involve 'observation, from a fixed roadside site, of "natural" (uninfluenced) driver behaviors and/or driver responses to traffic events (stimuli)'. These two types of situations were termed nonintervention and intervention, respectively.

Data obtained in *nonintervention situations* which would discriminate the alcohol-impaired driver would have greater utility for the potential user (e.g., police officers), since the motorist's driving behavior would not have to be manipulated by the observer in order to collect such data. This advantage over intervention situations is appealing both legally and logistically. Thus, for the nonintervention samples of driving behavior at the 'primary site', an electronic system was designed to sense remotely and then record both speed and lateral movement of cars traveling along straight segments of rural roadway.

The *intervention situations* were more challenging and have higher probable validity and payoff. We had originally designed a number of intervention situations based on the results of experiments which showed alcohol impairment of the ability to divide attention effectively. However, due to certain legal developments, it became absolutely impossible to use any of these situations on the public highways. Consequently, our closest approximation of an intervention situation was the so-called 'secondary site', at which drivers were directed to stop by a state police officer.

The data-recording system for secondary site performance measures was considerably simpler than that needed at the primary site. The performance measures of interest were speed and speed change as drivers came to a halt near the police officer. A doppler radar antenna was used to measure speed. Since the maximum range at the secondary site was only about 500 ft (152 m), no signal amplification was required and the signal could be recorded on a standard cassette tape recorder. The necessary code pulses were entered in the computer as the speed signal was processed by the PDP-12 computer.

Computer-generated graphs of speed at the secondary site, plotted as a function of stopping distance, were produced and analyzed visually. These plots of the changes in speed provide us with some very important data on actual stopping performance under alcohol. Since it is not feasible to present and to overlay in this volume the transparencies of the computer plots projected at the Symposium, a sample of two low and two high BAC drivers has been selected to serve as illustrative examples. Smoothed representations of their actual computer-generated plots (as presented in Helsinki) have been prepared and are presented as the four stylized curves in figure 3.

The plots are intriguing because they exhibit different patterns of stopping behavior. More specifically, drivers with zero or very low BACs generally stopped in a relatively smooth and consistent manner (figure 3a, b), whereas

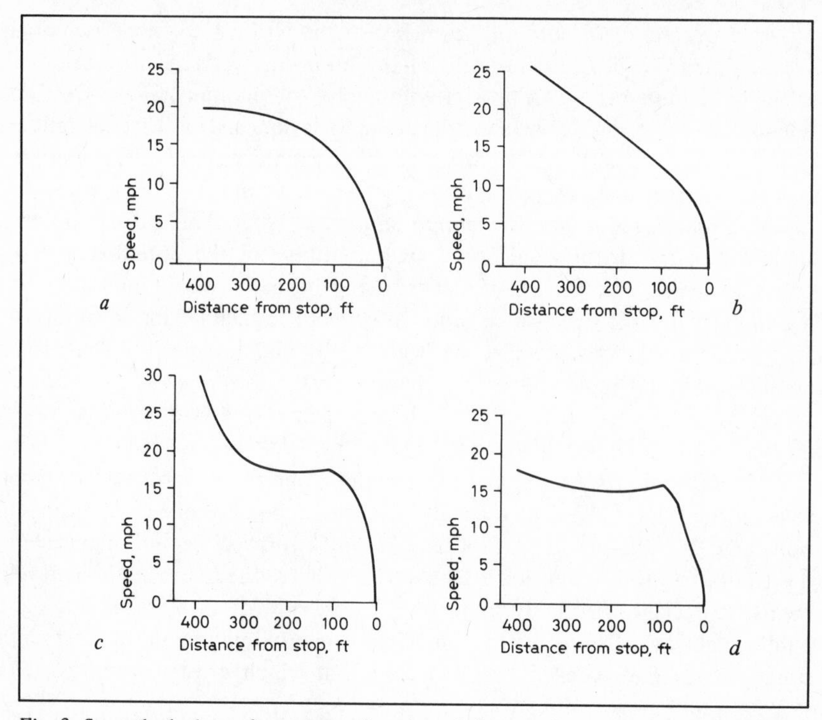

Fig. 3. Smoothed plots of speed as a function of distance from stopping point for four motorists at two roadside surveys. *a* BAC, 11 mg%; roadside survey number (RB), 32; car number (C), 93. *b* BAC, 30 mg%; RB, 16; C, 10. *c* BAC, 130 mg%; RB, 32; C, 89. *d* BAC, 120 mg%; RB, 16; C, 11.

drivers with medium to high BACs manifested less consistent behaviors (fig. 3c, d).

After visually examining a number of these plots of stopping behavior, one is impressed by the differences in patterns between drunk and sober drivers on exactly the same strip of roadway driven under the same environmental conditions on the same night (fig. 3: c versus a, and d versus b). However, perhaps even more impressive in terms of the implications for research and enforcement is the similarity between these two types of drivers at *different* sites. If we examine the plots of stopping behavior from drivers with zero or low BACs obtained at different places in the state on different nights and under different conditions, we find a striking congruence in the braking pattern (fig. 3a, b). Even more interesting, however, is the striking congruence obtained by comparing the plots of several high BAC drivers obtained at different places on different nights and under different environmental conditions (fig. 3c, d). It can be seen by inspecting these plots from high BAC drivers (fig. 3 c, d) that despite some possible differences in deceleration at the beginning of the measurements, the deceleration during the final 100 ft (30 m) was very rapid, which indicates a more abrupt stop with higher *g* forces.

Discussion

In comparing these 'natural' on-road stopping patterns with those obtained in our simulated passing experiment (16, 33, ch. 8), some interesting similarities appear in terms of alcohol influences upon braking performance. In the experiment, for example, the high BAC subjects did not begin to apply the brakes as quickly as they did when sober; however, when they did, they exerted more pressure faster than when sober and also completed the stop in less total time less smoothly at high BACs than they were with no alcohol. (Stopping accuracy was not a consideration in this experiment.) Accordingly, the stopping patterns of these well-practiced experimental subjects were congruent with the stopping patterns obtained from high BAC motorists (fig. 3 c, d) on the public highways who were simply responding to the perhaps unexpected, but not unusual appearance of a police officer with his car and flashing beacon light signalling the motorist to stop.

To extrapolate a bit beyond these data, it does not seem unreasonable to assume that this type of abrupt braking performance observed at high BACs is less controlled and thus more likely to lead to skidding and crashing than is the more gradual and smooth deceleration observed in the cases of motorists and experimental subjects with zero BACs. In other words, *the actual braking responses made under high BACs are conducive to losing control of the car.*

Thus, regarding the information-processing sequence discussed above, there is evidence that at least in terms of quality (e.g., brake-pressure modulation), alcohol impairs the final stage, that is, the response output or implementation stage. One could hypothesize that the lower quality, abrupt braking performance observed at high BACs in the two studies discussed above results from less time remaining available to the high BAC driver for stopping because alcohol has increased the information-processing time at some previous stage. The best evidence we have available at the moment would indicate that the greatest delay occurs at the stage of response encoding or *response selection.*

Since reaction times show relatively little alcohol impairment in tasks that involve high stimulus-response association strength and that require straightforward and highly familiar responses, the braking response times in the simulated passing study (16) may have shown a relatively small amount of alcohol impairment because the braking task met these criteria and furthermore was very unambiguous. In a similar fashion, one could also reason that the appearance of a police officer signalling a motorist to stop is also a straightforward stimulus situation. In contrast, however, the brake response times in the divided attention experiments using the instrumented car revealed large and significant increases at high BACs, especially with extrafoveal stimulation.

Thus, taken in combination with the abrupt braking patterns observed at high BACs with both motorists and subjects, *it seems reasonable to assume that high BACs increase the time necessary to begin applying the brakes, as well as reduce the degree of control in the actual use or modulation of the brakes during the course of stopping.* In conjunction with alcohol impairment of time-sharing capacity, these two interrelated considerations would seem to provide the basic ingredients of the alcohol contribution to highway crashes. Perhaps we have thereby fashioned the first rough plank for bridging the gap between epidemiology and experimentation.

Summary

It is frequently observed that alcohol-impaired drivers involved in certain types of crashes apparently 'couldn't stop in time'. However accurate this observation, it is simply a description and is not an adequate explanation. Therein lies the basis for the gap between epidemiology and experimentation.

If alcohol actually does degrade a motorist's performance and increase the probability of his being responsible for a fatal crash, then alcohol-induced changes in driving behavior should be manifest and should be measurable. However, no controlled study has previously been conducted to obtain systematic but *unobtrusive* data on the actual influences of alcohol upon real-world driving behavior in its natural environment. One recently completed field study is reported which was designed to provide such data by means of unobtrusive electronic measures of nocturnal driving performance.

In the present paper, to illustrate one approach to closing the gap between epidemiology and experimentation, one combination variable was selected which lends itself readily to comparison across the full spectrum of alcohol investigation: *reaction time and braking performance.* A review of the literature concerning this interrelated variable examined alcohol influences upon reaction time as investigated in laboratory, simulator, and instrumented car experiments, as well as alcohol influences upon braking performance in instrumented car experiments and in our recent field study involving unobtrusive electronic measures.

The reviewed experiments were interpreted in terms of a conceptualization of the information-processing sequence which leads up to brake use. On the basis of the reviewed results, it was concluded that alcohol increases reaction time (both simple and choice) appreciably more in driving situations than in laboratory experiments. It was also concluded that a consistent alcohol impairment of the *qualitative* aspects of braking performance is manifest in driving situations, for example, as reflected by changes in brake-pressure modulation. More specifically, the braking performance of motorists and subjects at high BACs is abrupt, unsmooth, and less controlled than that of sober motorists or the same subjects with no alcohol.

It was concluded that at least in terms of quality, alcohol impairs response implementation of the final stage of the information-processing sequence. It was hypothesized that the abrupt, lower quality braking performance observed at high BACs results – at least in part – from less time remaining available for stopping because the information-processing time has been increased by alcohol at some previous stage – apparently the response-selection stage. At the most general level, it was suggested that high BACs both increase the time necessary to begin applying the brakes, as well as reduce the degree of control in the actual use of the brakes during the course of stopping. These two factors in combination probably account for a large part of the alcohol contribution to highway crashes.

References

1 *Barry, H., III:* Motivational and cognitive effects of alcohol. J. saf. Res. *5:* 200–221 (1973).

2 *Barry, H., III:* Motivational and cognitive effects of alcohol; in *Perrine* Alcohol, drugs, and driving. National Highway Traffic Safety Administration, Technical Report, DOT HS-801-096, 1974.

3 *Belt, B.L.:* Driver eye movements as a function of low alcohol concentrations. Driving Research Laboratory, Department of Industrial Engineering, Ohio State University (USPHS), 1969.

4 *Brown, I.D.:* Safer drivers. Br. J. Hosp. Med. *1970:* 441–450.

5 *Carpenter, J.A.:* Effects of alcohol on some psychological processes: a critical review with special reference to automobile driving skill. Q. J. Stud. Alcohol *23:* 274–314 (1962).

6 *Carr, B.; Borkenstein, R.F.; Perrine, M.W.; Van Berkom, L.C., and Voas, R.B.:* International conference on research methodology for roadside surveys of drinking-driving. NHTSA, Technical Report, DOT HS-801-220, 1974.

7 *Chastain, J.D.:* Effects of 10 % blood alcohol on driving ability. Traf. saf. *5:* 4–5 (1961).

8 *Damkot, D.K.; Perrine, M.W.; Whitmore, D.G.; Toussie, S.R., and Geller, H.A.:* On-the-road driving behavior and breath alcohol concentration, vol. I, II. NHTSA, Technical Report, DOT HS-364-3-757, 1975.

9 *Heimstra, N. and Struckman, D.:* The effects of alcohol on performance in driving simulators. Proc. TRANSPO 72 Conf. on Alcohol and Traffic Safety, Washington, 1972.

10 *Huntley, M.S., jr.:* Influences of alcohol and S-R uncertainty upon spatial localization of time. Psychopharmacologia *27:* 131–140 (1972).

11 *Huntley, M.S., jr.:* Alcohol influences upon closed-course driving performance. J. saf. Res. *5:* 149–164 (1973).
12 *Huntley, M.S., jr.:* Effects of alcohol and fixation-task difficulty on choice reaction time to extrafoveal stimulation. Q. J. Stud. Alcohol *34:* 89–103 (1973).
13 *Huntley, M.S., jr.:* Alcohol influences upon closed-course driving performance; in *Perrine* Alcohol, drugs, and driving. NHTSA, Technical Report, DOT HS-801-096, 1974.
14 *Huntley, M.S., jr.:* Effects of alcohol, uncertainty, and novelty upon response selection. Psychopharmacologia *39:* 259–266 (1974).
15 *Huntley, M.S., jr.; Kirk, R.S., and Perrine, M.W.:* Effects of alcohol upon nocturnal driving behavior, reaction time, and heart rate. Paper presented at SAE Automotive Engineering Congress, Detroit 1972.
16 *Huntley, M.S., jr.; Perrine, M.W., and Kirk, R.S.:* Influences of alcohol upon control-response times and brake-pressure modulation during simulated passing. Proc. 1st Int. Conf. on Driver Behaviour, Zurich 1973 (International Drivers' Behaviour Research Association, Courbevoie 1973).
17 *Hurst, P.M.:* Epidemiological aspects of alcohol in driver crashes and citations. J. saf. Res. *5:* 130–148 (1973).
18 *Hurst, P.M.:* Epidemiological aspects of alcohol in driver crashes and citations; in *Perrine* Alcohol, drugs, and driving. NHTSA, Technical Report, DOT HS-801-096, 1974.
19 *Levine, J.M.; Greenbaum, G.D., and Notkin, E.R.:* The effects of alcohol on human performance: a classification and integration of research findings (American Institute for Research, Washington 1973).
20 *Lovibond, S.H. and Bird, K.:* Effects of blood alcohol level on the driving behavior of competition and non-competition drivers. Proc. 29th Int. Congr. on Alcoholism and Drug Dependence, Sydney 1970.
21 *Michon, J.A.; Eernst, T.E.; Koutstaal, G.A.; Vunderink, R.A.F.; Spiekman, L.; Schipper, J.M., and Burrij, S.:* Het effect van kleine doses alcohol op enkele variabelen van het gedrag van de automibilist, door. Report No. 1ZF 1970E6, Institute for Perception RVO-TNO, Soesterberg 1970.
22 *Moskowitz, H.:* Laboratory studies of the effects of alcohol on some variables related to driving. J. saf. Res. *5:* 185–199 (1973).
23 *Moskowitz, H.:* Alcohol influences upon sensory motor-functions, visual perception, and attention; in *Perrine* Alcohol, drugs, and driving. NHTSA, Technical Report, DOT HS-801-096, 1974.
24 *Perrine, M.W.:* Alcohol influences upon driving-related behavior: A critical review of laboratory studies of neurophysiological, neuromuscular, and sensory activity. J. saf. Res. *5:* 165–184 (1973).
25 *Perrine, M.W.:* Alcohol and highway safety; in Alcohol and health: new knowledge. 2nd special report to the US Congress, Stock No. 1724–0399 (US Government Printing Office, Washington 1974).
26 *Perrine, M.W.:* Alcohol, drugs, and driving. NHTSA, Technical Report, DOT HS-801-096, 1974.
27 *Perrine, M.W.:* Alcohol experiments and driving-related behavior: a review of the 1972–73 literature – Alcohol Countermeasures Literature Review. NHTSA, Technical Report. DOT HS-801-266, 1974.
28 *Perrine, M.W.:* Alcohol influences upon driving-related behavior: a critical review of laboratory studies of neurophysiological, neuromuscular, and sensory activity; in *Perrine* Alcohol, drugs, and driving. NHTSA, Technical Report, DOT HS-801-096, 1974.

29 *Perrine, M.W.:* Alcohol experiments and driving-related behavior: a review of the 1974 literature. Committee on Alcohol and Drugs, National Safety Council, 1975.
30 *Perrine, M.W.:* Alcohol involvement in highway crashes: a review of the epidemiologic evidence; in *Schultz* Clin. Plas. Surg, vol. 2, pp. 11–34 (Saunders, Philadelphia 1975).
31 *Perrine, M.W.; Damkot, D.K., and Worden, J.K.:* Roadside research (a documentary color film), Project ABETS, University of Vermont, 1975.
32 *Perrine, M.W. and Huntley, M.S., jr.:* Influences of alcohol upon driving behavior in an instrumented car. NHTSA, Technical Report, DOT HS-800-625, 1971.
33 *Perrine, M.W. and Huntley, M.S., jr.:* Alcohol and driving behavior: experiments with instrumented cars on closed courses. NHTSA, Technical Report, DOT FH-11-7469, 1975.
34 *Smith, E.E.:* Choice reaction time: an analysis of the major theoretical positions. Psychol. Bull. *69 (2):* 77–110 (1968).
35 *Stroh, C.M.:* Roadside surveys of drinking-driving behaviour. Proc. Conf. on Medical, Human and Related Factors Causing Traffic Accidents, Including Alcohol and Other Drugs, Ottawa, Ont. Traffic Injury Research Foundation of Canada, 1972.
36 *Stroh, C.M.:* Alcohol and highway safety; roadside surveys of drinking-driving behavior: a review of the literature and a recommended methodology. Road and Motor Vehicle Safety Traffic, Ministry of Transport, Ontario, CTS-lb-74, 1974.
37 *Tharp, V.K., jr.; Rundell, O.H., jr.; Lester, B.K., and Williams, H.L.:* Alcohol and information processing. Psychopharmacologia *40:* 33–52 (1974).
38 *Wallgren, H. and Barry, H., III:* Actions of alcohol (Elsevier, New York 1970).
39 *Zylman, R.:* Analysis of studies comparing collision-involved drivers and non-involved drivers. J. saf. Res. *3:* 116–128 (1971).
40 *Zylman, R.:* A critical evaluation of the literature on 'alcohol involvement' in highway deaths. Accid. Anal. Prev. *6 (2):* 163–204 (1974).

Prof. *M.W. Perrine,* Project ABETS, Psychology Department, John Dewey Hall, *Burlington, VT 05401* (USA)

Mod. Probl. Pharmacopsych., vol. 11, pp. 42–45 (Karger, Basel 1976)

Practical Aspects of the Routine Measurement of Alcohol and Drugs in Drivers

A Preliminary Report

A. Alha, R. Honkanen, M. Karlsson, K. Laiho, M. Linnoila and I. Lukkari
Department of Forensic Medicine, University of Helsinki, Helsinki

As was demonstrated by *Borkenstein* 10 years ago, the increase in the accident risk in traffic attributable to alcohol is well established. A similar concentration-related increase of the risk of pedestrians falling as a consequence of alcohol has been demonstrated to this symposium by *Honkanen et al.* With a view to the improvement of accident statistics, and to justification of their use in evaluation of the effects of alcohol and drugs, and for possible legislative measures, the availability and the accuracy of the techniques applied in the measurement of alcohol and drugs in body fluids must be satisfactory. Moreover, these techniques should be sufficiently rapid in application to enable their use during night razzias by the police.

Experimental

Alcohol. Routine measurement of the blood alcohol concentrations (BAC) has been made in the Department of Forensic Medicine by simultaneous application of the Widmark, alcohol dehydrogenase (ADH), and gas chromatographic methods. In November 1974, and in January–February 1975, we tested the suitability of the quantitative measurement of alcohol in expired air to indicate the BAC. Parallel estimations of the BAC were made from the breath air by the ASD instrument (US Department of Transportation), based upon electronic oxidation, and from blood samples by means of the Widmark and ADH methods. The results of tests made on 44 persons suspected of drunken driving and examined during 1 weekend period of duty will be presented. Calibration control of the ASD instrument used by means of the Decatur Alcohol Breath Simulator, and using 1 mg/g alcohol concentration, was proved to be correct.

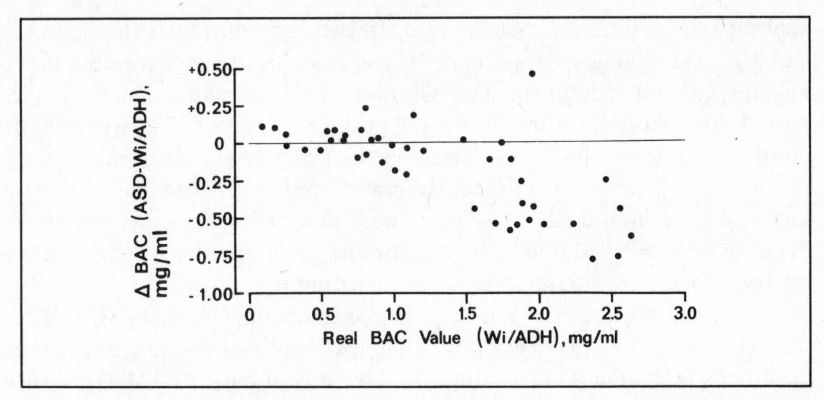

Fig. 1. Correlation of the actual blood alcohol concentrations (BAC) measured by Widmark and alcohol dehydrogenase methods and the BAC estimates calculated by the ASD instrument from the breath air. Δ BAC represents the difference, positive or negative, of the ASD reading and the corresponding true BAC value (mg/ml = ‰).

The results obtained are given in figure 1. It is apparent that if the BAC was 1.5 mg/ml or more (up to 2.87 mg/ml), the breath alcohol content, as measured by the ASD instrument, in nearly every case underestimated BAC showing too low values for BAC in comparison with the levels actually measured in the blood. The negative difference was 0.79 mg/ml at its highest. At the BACs of 1 mg/ml and less, the indication given by the breath air measurement was close to the actual BACs. However, at the BACs of less than 0.5 mg/ml, the ASD tended to overestimate the blood alcohol, as was apparent in the slightly higher BACs derived by this method.

For further study of this problem, three test subjects were given alcohol (1.8 g/kg absolute alcohol), ingested within the course of about 2.5 h. Calibrations of the ASD instrument were made by means of 1 and 2 mg/ml alcohol concentrations. Parallel ASD (with the simultaneous employment of three items of equipment) and blood alcohol tests were made at intervals of 1 h from the beginning of the experiment, five times in all. BAC maxima of 1.71–1.81 mg/ml were recorded after 3 h. No indication was found of the divergence between real BAC values and the ASD data mentioned above.

Drugs in urine and plasma. In spring 1974, persons arrested by the police on suspicion of driving under the influence of alcohol or/and drugs were subjected to drug analysis, with urine taken from 100 persons in all parts of Finland. In April 1975, plasma taken from 100 successive cases of suspected drunken-driving, was examined in our department.

Urine (25–50 ml) was ether-extracted successively to acid and neutral, and to alkaline groups. Benzodiazepines were hydrolysed to the corresponding benzophenones. Qualitative detection was effected by 5 thin-layer chromatography and 3 gas chromatography systems. Of 100 cases investigated in spring 1974, 24 samples did not exhibit alcohol. Of these alcohol-negative cases, 18 (75 %) proved drug-positive. Of the 76 alcohol-positive cases 25 (33 %) were also drug-positive. The drug found qualitatively in the majority (77 %) of this total of 43 drug-positive samples was a benzodiazepine derivative, in most cases diazepam or lorazepam. In individual cases, there were detected some other psychotropic drugs, hypnotics or sedatives (chlorprothixene, levomepromazine, meprobamate, methaqualone, barbiturates, primidone, ephedrine).

For isolation from the *plasma,* n-heptane + isoamyl alcohol (98.5 + 1.5 v/v) was used as solvent (5). Two sets of plasma (4 ml each) were extracted under conditions that were acidic (2 drops of 2 *N* HCl) or basic (2 g MgO), with 10 + 5 ml of the solvent, and with the extracts being filtered and evaporated to 1 ml at 40 °C. 1- to 5-μl aliquots of the filtrates were injected into gas chromatograph: Varian 2700 Aerograph, col 1.5 % OV-17 on Gas-Chrom Q, system 1 AFID programmed at 100–260 °C for acid, neutral and basic drugs in general, system 2 ECD, at 270 °C for benzodiazepines, and system 3 Perkin-Elmer F 11, col. 5 % Carbovax + 5 % KOH on Chromosorb G, FID at 160 and 180 °C for amphetamines. Recoveries were calculated from the peak heights of 2–4 analyses. The external standards used were flurazepam for system 2, and diethylaniline for system 3. Thin-layer chromatography was used for confirmation. Urine (available in 65 cases) was also examined, along with special assays for salicylates (4) and isoniazid (1) on plasma, and for morphine on urine (2).

Alcohol was found in 98 cases. Drugs were detected in 5 cases, in 4 cases combined with alcohol. In plasma, dipotassium chlorazepate 1.6 μg/ml, diazepam 0.2 μg/ml + desmethyldiazepam 0.3 μg/ml, carbromal 16 μg/ml, or carbutamide traces were found. Examination of urine also revealed meprobamate in the carbutamide case, and traces of phenobarbitone in one other case.

Discussion

In regard to the measurement of BACs, traditional laboratory methods provide accurate results which can be relied upon for legislative measures if a punishable limit of BAC is to be instituted. Quantitative instant tests made on the road by a police officer carrying an ASD or other instrument with him call for careful calibration of the instrument and a power supply. However, the results obtained may still prove too approximate. To some extent, our results agree with those reported by Dr. *Jones* to this meeting, with the statement that the ASD instrument overestimates the true BAC at low alcohol levels, and underestimates it at high alcohol levels. *Schmutte et al.* (3) have arrived at comparable results in the assaying of BAC from the breath with an Alcolyser operating by gas chromatography, and from blood samples by gas chromatography. In their work, the number of cases with negative difference shown by the breath method increased in stress and hyperventilation within the range of 0.4–1.9 mg/ml. They concluded that the difference might be attributable to physiological fluctuations in breath alcohol content and breath sampling. We are inclined to suggest that in drunken-driving cases with high BACs, some biological effect of alcohol can contribute to the discrepancy found, which could be a change in the ventilation perfusion ratio in the lungs. But proportion of this descrepancy is obviously attributable to inaccurate functioning of the ASD instrument, as is evident from the overestimation of blood alcohol at low BACs found by Dr. *Jones,* and in our experiments.

The detection of drugs in drunken-driving cases may be complicated. A combined blood/urine analysis, with different methods for isolation and detection, is of value, particularly when the case report gives no hint of the drugs concerned, and the specimens are minute. Urine is not always available. When, however, drug analysis is an urgent requirement, a plentiful blood sample (20 ml or more) is of highest importance. The results obtained indicate that at present, driving under the influence of drugs is a minor problem here as compared with the dangers arising from alcohol, tiredness, a disturbed state of mind, excessive speed, or other human factors.

References

1 *Deeb, E.N. and Vitagliano, G.R.:* The blood level determination of PAS and INH with vanillin. J. Am. pharm. Ass. *44:* 182–185 (1955).

2 *Goldbaum, L.R.; Santinga, P., and Dominiguez, A.M.:* A procedure for the rapid analysis of large numbers of urine samples for drugs. Clin. Toxicol. *5:* 369–379 (1972).

3 *Schmutte, P.; Strohmeyer, H. und Naeve, W.:* Vergleichende Untersuchungen von Atem- und Blutalkoholkonzentration nach körperlicher Belastung und besonderer Atemtechnik (Hyperventilation). Blutalkohol *10:* 34–42 (1973).

4 *Trinder, P.:* Rapid determination of salicylate in biological fluids. Biochem. J. *57:* 301–303 (1954).

5 *Zingales, I.A.:* Determination of chlordiazepoxide plasma concentrations by electron capture gas-liquid chromatography. J. Chromat. *61:* 237–252 (1971).

Prof. *A. Alha,* Department of Forensic Medicine, University of Helsinki, *Helsinki* (Finland)

Mod. Probl. Pharmacopsych., vol. 11, pp. 46–50 (Karger, Basel 1976)

A Case-Control Study on Alcohol as a Risk Factor in Pedestrian Accidents

A Preliminary Report[1]

R. Honkanen, L. Ertama, P. Kuosmanen, M. Linnoila and T. Visuri[2]
Department of Public Health Science, University of Helsinki, Helsinki

It is well known that blood alcohol concentrations (BAC) of over 0.5 g/l increase a driver's risk of getting involved in traffic accidents and that this risk rises steeply with increasing BACs (2). Alcohol as a risk factor in other accidents has been given scanty attention. *Haddon et al.* (4) found that fatally injured pedestrians hit by motor vehicles had alcohol in the blood more often than site-matched controls. There are also studies that show more frequent alcohol involvement in accidents than in other emergencies (6). However, we do not yet know how alcohol increases the risk of involvement in nontraffic accidents.

Accidental falls are the most common type of pedestrian accidents, and they account for more than one third of the injuries treated at our emergency stations. They are of special importance in Finland because of the long winters with snow and slippery roads. Experimental studies (3) and common sense support the view that alcohol increases the risk of falling, but controlled population studies are needed. However, the control is quite difficult to arrange (5).

The purpose of the present work was to obtain information on the role of alcohol as a risk factor in pedestrian accidents. It was noted that with regard to falls in public places, the control group could be formed in the same way as was done by *Haddon et al.* (4) in their study on pedestrian fatalities.

Materials and Methods

The study was made during two 5-week periods in 1975, the first period (3 Jan.–6 Feb.) representing winter and the second (5 May–8 June) summer conditions.

[1] Supported by a grant from the Yrjö Jahnsson Foundation, Helsinki.
[2] The authors are grateful to the US Department of Transportation for the ASD-meters.

Cases. The cases were the 313 adults (15 years of age or older) who fell in public places and the 28 adults who were hit by motor vehicles in the city of Helsinki between 3 p.m. and 11 p.m. and who arrived by 11 p.m. on the same day at the emergency station of the Department of Orthopaedics and Traumatology, Helsinki University Central Hospital. The cases were interviewed by the authors at the emergency station, after which the blood specimens for alcohol determination were taken. The interview included all the questions set to the controls (see following) as well as questions concerning the accident. Alcohol concentrations were measured from serum by gas chromatography in the Department of Public Health Science, University of Helsinki. Propanol in 0.01 % solution was used as the internal standard. The results were reported as grams of ethanol per litre of whole blood.

No specimen could be obtained from 14 cases. Half of them refused to cooperate, and the rest were missed by the interviewers. Five of them were drunk.

Controls. The control group was formed by visiting each accident site exactly *1 week after* the accident and by interviewing the first 2 adult pedestrians of the same sex as the patient reaching the site. All the controls were interviewed within 30 min before or 45 min after the hour of the accident. The authors themselves stopped the control subjects. Precautions were taken to ensure random selection. *The interview included* questions about the following matters: place of residence, walking activity, age, marital status, occupation, walking ability, smoking, last intake of medicine and alcohol. In addition, signs of alcohol use, type of shoes and sole material as well as weather conditions were recorded. After the last question, the control subject was asked to blow into one of the ASD-breathalyzers (US Government, Dept. of Transportation). The full cooperation in 91 % of the control persons was obtained. The rest either refused to stop (5 %) or did not blow because of refusal (2 %) or inability (2 %). One of the 35 who refused to stop and 7 of the 27 who consented to be interviewed but did not register a BAC value were intoxicated.

Results

Out of total 341 cases only 28 were hit by motor vehicles. A great majority (231/341) of the patients fell on the same level, slipping being the most usual injury mechanism in winter and stumbling in summer. The accident cases were

Table I. Percent distribution of cases by BAC in both sexes in winter and summer

BAC, g/l[1]	Males			Females		
	winter (n = 81)	summer (n = 95)	total (n = 176)	winter (n = 86)	summer (n = 62)	total (n = 151)
0–0.5	27	17	22	84	76	81
0.6–	73	83	78	16	24	19
Total, %	100	100	100	100	100	100

[1] Excluded were 14 cases with BACs unknown.

Table II. Distribution of cases and controls by BAC in falls and in those hit by motor vehicles

BAC, g/l	Falls		Motor vehicles	
	cases	controls	cases	controls
0	126	118	13	12
0.1	11	282	0	31
0.2	3	106	1	8
0.3	2	21	0	0
0.4–0.5	3	15	1	1
0.6–1.0	10	20	2	0
1.1–1.5	29	16	7	3
1.6–2.0	35	10	2	0
2.1–2.5	39	7	0	0
2.6–3.0	26	2	0	0
3.1–	17	1	0	0
Unknown	12	28	2	1
Total	313	626	28	56

considerably older than the controls, the mean age being 44.9 years in the former and 37.4 years in the latter group. This difference was more marked in the females than the males. The total number of cases was smaller in summer than in winter (table I) and the difference was significant among the females ($p < 0.01$). A majority of these accidents occurred in the afternoon rather than in the late evening (55 % between 3 and 6 p.m.).

Significant alcohol involvement (BAC over 0.5 g/l) was found in 78 % of the males and 19 % of the females (table I). In both sex groups it was more frequent in summer than in winter. Table II presents the distribution of cases and controls by BAC. The low BACs in the control series show that the most usual negative reading was 0.1 g/l, the next two being zero and 0.2 g/l. A profound difference in alcohol involvement between cases and controls was found: it was 53 % among the cases and 15 % among the controls ($p < 0.001$). The relative risk (RR) of an injury from falling at each BAC (fig. 1) shows a decrease at the low BACs (0.3–0.5 g/l), slight increase at the BAC of 0.6–1.0 g/l and a rapid rise thereafter, being 29 at BACs of over 2.0 g/l.

In spite of the small number of cases hit by motor vehicles the difference in alcohol involvement between cases and controls was highly significant ($p < 0.001$) also in this category. The risk of being hit was 10 times higher in those with alcohol involvement than in those without it.

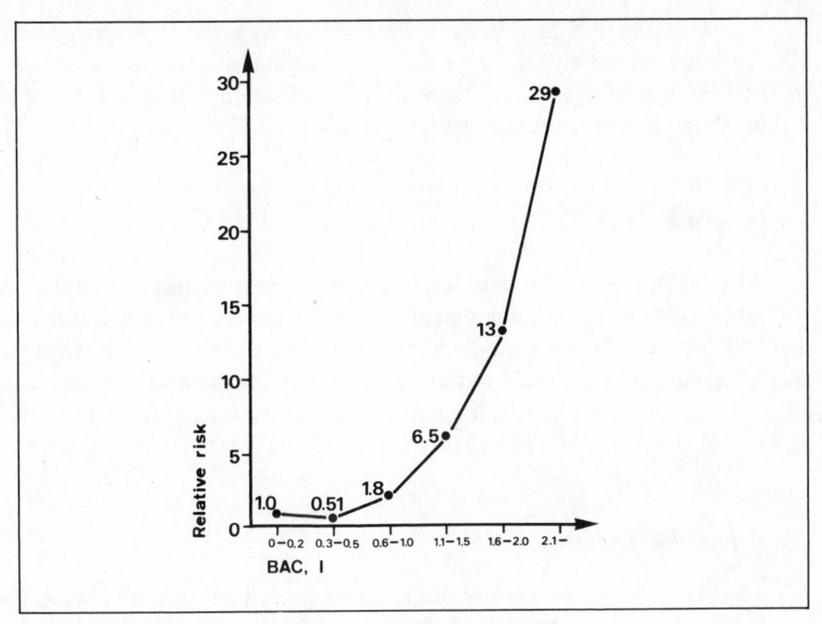

Fig. 1. Relative risk of injury from accidental falls by BAC.

Discussion

The results indicate that injuries from pedestrian accidents were not frequent among healthy and sober persons. The two outstanding groups of patients were: (1) young and middle-aged males who were involved in accidents while intoxicated and (2) middle-aged and old females who fell in winter because of slipping or were hit by motor vehicles because of a physical or mental handicap.

The primary purpose of this study, the estimation of the risk of injury from falling at different BACs, was achieved. The increase of the RR was so marked that it cannot be caused by chance or by methodological weaknesses. One of these weaknesses was that different methods of the BAC determination were used in the case and control series: gas chromatography used in the case series gives reasonably quantitative results, but the semiquantitative ASDs used in the control series tend to give too high readings at low BACs and too low readings at high BACs (1). This is liable to distort the RR values, namely by decreasing it at low (0.1–0.5) and moderate (1.1–2.0) BACs, but increasing it at BACs of over 2.0 g/l. The good response rate achieved prevents any serious bias in this respect.

That the RR in this study did not increase so rapidly as it did in Borkenstein's study might be due to the fact that driving skill deteriorates at lower BACs and more markedly than does walking.

According to this investigation alcohol is a high-risk factor in pedestrian accidents and apparently causes a remarkable portion of them. Age seems also to increase the risk of injury from these accidents. Therefore, the age factor should be considered in more exact estimations of risk.

Summary

Alcohol as a risk factor in pedestrian accidents was studied so that the BACs were measured by gas chromatography in 341 accident patients and by ASD-breathalyzers in 682 sex-, site-, and time-matched pedestrians not involved in accidents. Alcohol was found in the blood in 53 % of the accident cases and in 15 % of the controls. The RR, if 1.0 at zero BAC, was 0.59 at BAC 0.3–0.5, 2.2 at 0.6–1.0, 6.8 at 1.1–1.5, 13 at 1.6–2.0, and 29 at BACs of over 2.0 g/l.

References

1 *Alha, A.; Laiho, K., and Linnoila, M.:* Blood alcohol in breath determined by ASD and in blood by Widmark and ADH-methods. A comparative study. Blutalkohol (in press).

2 *Borkenstein, R.F.; Crowther, R.F.; Shumate, R.P.; Ziel, W.B., and Zylman, R.:* The role of a drinking driver in traffic accidents. Department of Police Administration, Indiana University (1964).

3 *Goldberg, L.:* Quantitative studies on alcohol tolerance in man. The influence of ethyl alcohol on sensory, motor and psychological functions referred to blood alcohol in normal and habituated individuals. Acta physiol. scand. *5:* suppl. 16, pp. 1–128 (1943).

4 *Haddon, W., jr.; Valien, P.; McCarroll, J.R., and Umberger, C.J.:* A controlled investigation of the characteristics of adult pedestrians fatally injured by motor vehicles in Manhattan. J. chron. Dis. *14:* 655–678 (1961).

5 *Waller, J.A.:* Personal commun.

6 *Wechsler, H.; Kasey, E.; Denise, T., and Demone, H.:* Alcohol level and home accidents. Publ. Hlth Rep., Wash. *84:* 1043–1050 (1969).

Dr. *R. Honkanen,* Department of Public Health Science, University of Helsinki, *Helsinki* (Finland)

Mod. Probl. Pharmacopsych., vol. 11, pp. 51–56 (Karger, Basel 1976)

Traffic Accidents and Psychomotor Test Performance

A Follow-Up Study[1]

Sauli Häkkinen
Helsinki University of Technology, Espoo

An essential problem in the methodology of the research on drugs and driving is the question of criterion and validity. It is not possible to perform studies of this kind in real traffic circumstances. Simulation and different kinds of psychological tests have to be resorted to. But how much do we know about the validity of these methods?

We may accept the view that accidents are the most important basic criterion for traffic studies. Accidents are, however, a typical stochastic variable with a rare occurrence. Therefore, we run into difficulties as we are not able to control several changing variables in traffic. A very typical question in this field is the effect of individual differences on accident rates or so-called accident proneness. Much research, discussion and argumentation on this concept has occurred during the last 50 years, a characteristic indication of this being the book of nearly 500 pages by *Shaw and Sichel* entitled Accident Proneness (1971).

One of the relatively rare studies in which the stable nature of individual differences in accident rates has been shown was made in Helsinki in 1958 (*Häkkinen,* 1958). In this basic study the reliability of bus and tram driver accidents was examined by using the correlations between the accidents incurred during two consecutive time periods, by analyzing the distributions of the accidents, and by comparing the group averages. Correlations between accidents in two consecutive years were of the order of 0.20–0.35 and between two consecutive 4-year periods of the order of 0.60–0.70. This means that if accidents are used as an individual criterion of accident proneness, the exposure time must be several years in homogenous traffic circumstances.

[1] This research was partially supported by the Central Organisation for Traffic Safety in Finland.

The basic test psychological study was carried out with 100 drivers. The period of exposure was 8 years and the reliability of accidents was 0.80. 14 tests with a total of 300 variables were used. A comparison of the averages for the safe and the accident drivers was made and correlations and factor analysis were computed for gaining a concise description of the safe and the accident-prone type of drivers. No significant differences were found between the safe and accident groups in intelligence tests or in the variables measuring simple, disjunctive or choice reaction times.

In eye-hand coordination the accident group did worse, as a rule, than the safe group. The correlations between coordination variables and accident coefficient ranged from 0.10 to 0.35. The accident coefficient correlated highest with the variables of the psychomotor tests of a more complicated and special nature. In the Driving Apparatus Test (DAT) the subject had to respond to four different kinds of stimuli with specified hand and foot movements at the same time as he had to keep a pointer, which was moved by means of a steering wheel, on a winding 'highway'. Test subject's driving experience did not substantially modify his performance on this test.

The Expectancy Reaction Test (ERT) was a visual disjunctive reaction test in which the subject had to respond to stimuli of one kind only. The situation was made more difficult by means of distractions included in the test or external to it. In the DAT and in the ERT several variables were discovered which significantly distinguished the accident group from the safe one. These variables measured correct responses, errors and superfluous reactions. The highest correlations between the accident coefficient and these variables were 0.30–0.42.

In the Ambiguous Situation Test (AST) easy, difficult, and ambiguous situations were involved. The subject was required to respond to alternating light stimuli by turning a rod in specified directions. All movements of the rod were registered. The test variables pertaining to the ability domain (reaction time, speed, correct reactions, etc.) resulted in only a few differences between the groups. On the other hand, motor disturbances were more numerous in the accident than in the safe group, and, as the situation became more difficult, these disturbances increased more rapidly in the former than in the latter group. The highest correlations with the accident criterion were 0.30–0.43.

The factor analysis of 28 variables yielded six factors, which may be named as follows: intelligence, attention, coordination, simple reaction time, involuntary control of motor functions, and stability of behaviour. The accident coefficient was most strongly saturated (0.52) on the attention factor, as determined, primarily, by the correct responses in the DAT and ERT. The next highest loading of the accident variable was in the involuntary control of motor function (0.47). The variables descriptive of poor control were indicative of hastiness, susceptibility to disturbances, and motor restlessness. The loading of the criterion variable in coordination was relatively low (0.32) and in intelligence as well as in simple reaction time it was quite close to zero.

A part of the driver group considered in the basic study continued working in the same Municipal Transport Company. In the follow-up study the exposure time and accident figures for these drivers were collected from the whole period of driving in this company.

The shortest follow-up exposure time was 2.5 years and the longest was 18.5 years, the mean being 9.3 years. The total exposure time varied from 10.5 to 26.5 years with a mean of 17 years. The testing of the drivers took place during the last years of the basic exposure period. 66 out of the original group of 100 drivers belonged to this follow-up group.

The correlation between the accident coefficient (the number of accidents per year of exposure) of the first (basic) 8-year period and the second period of an average of 9 years was 0.56 corresponding to the reliability 0.72 for the total exposure period. This means that the accident behaviour of professional city drivers was highly constant over a period of driving of more than 20 years. We have at least in this case a very reliable accident criterion for studying individual differences in accident proneness. The large test battery offers a good possibility to compare individual test performances with accident behaviour over the whole exposure period and over various parts of it.

The highest correlations between the accident coefficient and the test variables for the original group of 100 drivers, and for the follow-up group of 66 drivers in different exposure periods are given in table I. The table reveals that the correlations of the test variables with the accident criterion in different exposure periods are of nearly the same order of magnitude. Correlations in the follow-up period are approximately equal to those in the first period, although the time lapse between the testing and the latter exposure period varied from 1 to 20 years. When the accident coefficient for the whole period was used as a criterion the validity correlations were as high as or higher than those for the original period. The means of the ten highest correlations (according to the basic study) were in the basic, follow-up and whole periods 0.38, 0.34, and 0.40, respectively.

Multiple regression analysis was used to calculate the highest correlation between the accident coefficient and the combined result of 18 test variables. Table II contains results of the regression analyses for different exposure periods when 5, 10 and 18 test variables were used for calculation. The multiple correlation varies from 0.70 to 0.80, which means that 50–65 % of the total variance of accidents has been explained by the test variables used. Five test variables already give a rather good validity, and the correlations increase very little after the inclusion of 10 test variables. The explanation for the follow-up period is higher than or equal to that for the basic period. The explanatory power is highest for the total period.

Table I. Highest correlations of accident coefficient to test variables in different exposure periods

Test variable	Basic study (n = 100)	Follow-up study (n = 66)		Total
	basic	basic	follow-up	
AST, motor disturbances I	0.43	0.49	0.43	0.50
DrT, correct reactions	0.42	0.49	0.41	0.48
DrT, simple steering	0.34	0.47	0.57	0.61
Rating of behaviour	0.34	0.26	0.17	0.22
ERT, superfluous reactions	0.32	0.42	0.43	0.48
ERT, correct reactions I	0.32	0.40	0.32	0.39
ERT, correct reactions III	0.32	0.29	0.35	0.37
DrT, missed signals	0.32	0.41	0.34	0.42
AST, motor disturbances II	0.27	0.28	0.25	0.29
DrT, errors	0.26	0.33	0.16	0.23
Mean of correlations	0.30	0.38	0.34	0.40

Table II. Multiple correlations between accident coefficient and test variables

Combinations of variables	Exposure period					
	basic study		follow-up study		total	
	r	% explained	r	% explained	r	% explained
5 test variables	0.71	0.51	0.71	0.50	0.75	0.56
10 test variables	0.74	0.55	0.76	0.58	0.80	0.64
18 test variables	0.75	0.56	0.77	0.60	0.81	0.65

The test variables included in the regression model were the same that are seen in the table of the highest validity correlations. The regression models for different exposure periods are quite similar. This means that no big changes had occurred in the personality factors affecting a safe or accident behaviour in traffic during 20 years. Similar results were gained in using factor analyses for different exposure periods.

One possible way of getting a concrete picture of what is the prediction power of the test battery, of gaining knowledge of the accident behaviour of the first exposure period, is to use fictional selection procedures and examine their

Table III. Results of fictional selection procedures

Accident behaviour	Accidents during follow-up period		
	basic	follow-up	
Better half of drivers	184	257	difference
Worse half of drivers	522	644	387 = 43 %
Total	706	901	
Sum of five test variables			
Better half of drivers	247	232	difference
Worse half of drivers	459	669	437 = 48.5 %
Total	706	901	
	Accidents in total period		
Sum of five test variables			
Better half of drivers	479	difference	
Worse half of drivers	1,128	649 = 40.4 %	
Total	1,607		

effect on the accident figures. The total number of accidents substained by the 66 drivers was 1,607 during the whole (average) exposure period of 17.3 years (table III).

The whole group of drivers was first divided into two equal groups on the basis of the accident figures in the first exposure period (184 and 522). Had the drivers been selected for the second (follow-up) period on the basis of these figures and had the average accident rate for the selected drivers been the same as in the better half of the total group, accidents would have numbered 387, or 43 %, less than they did in reality.

Had the basis of selection been the sum of the best five test variables and had the same procedure been used, there would have been 48.5 %, or 437, accidents less than in reality. This means that the prediction power of these five test variables was better than that of the information on accident figures for the first 8 years. The actual accident behaviour changed more than the 'potential' accident behaviour measured by five test variables in the early stage of driver experience. The correlations between the accident coefficient and the sum of five test variables for the basic, follow-up and the total period are 0.60, 0.60, and 0.67, respectively.

If the same selection procedure had been used before hiring these drivers and the worse half of the drivers had been replaced by the better half on the basis of five test variables, the number of accidents could have been reduced by 649 or by 40.4 %. These figures reveal the practical value of the test battery used.

In summary, the accident behaviour of professional city drivers proved to be very constant, and this behaviour is capable of being detected by specially planned psychological tests. These may provide an expedient for many kinds of research of driver behaviour not using actual driving variables – such as accidents – as a criterion.

References

Häkkinen, S.: Traffic accidents and driver characteristics. Finland's Institute of Technology, Scientific Researches No. 13, Helsinki (1958).

Shaw, L. and Sichel, H.: Accident proneness (Pergamon Press, Oxford 1971).

Prof. *S. Häkkinen,* Department of Mechanical Engineering, Helsinki University of Technology, *Otaniemi* (Finland)

Mod. Probl. Pharmacopsych., vol. 11, pp. 57–67 (Karger, Basel 1976)

Relation between Drug-Induced Central Nervous System Effects and Plasma Levels of Diazepam in Man

J. Orr, P. Dussault, C. Chappel, L. Goldberg and G. Reggiani
Bio-Research Laboratories, Pointe-Claire, Que., and Department of Alcohol Research, Karolinska Institute, Stockholm

The elucidation of a possible relationship between drug-induced central nervous system (CNS) effects and plasma levels of drugs is important as a basis for the understanding of changes in drug effects, of explaining inter- and intra-individual variability, of understanding variations in drug uptake, distribution and elimination and of elucidating underlying sites and mechanisms of action. This applies to the whole field of psychoactive drugs.

In the alcohol field there exists a great number of studies elucidating the close relationship between alcohol-induced CNS effects and blood ethanol levels as well as the influence of factors such as variations in beverage compositions, food intake, psychophysical relationships, the emergence of tolerance and post-drug (hangover) phenomena (for pertinent reviews see 1–5, 11). Of psychoactive drugs other than alcohol very few studies exist, where a significant relationship between biological effects and drug levels, based on adequate experimental data, has been demonstrated, even if a number of studies have attempted to show at least some relation to dose levels by comparing effects at a few time-points against dosage (6–9).

The growing importance of problems of bioavailability in the drug field also points to the need for evaluation and reevaluation of the problem of dose-response relationships. However, in order to elucidate these relationships, four conditions have to be fulfilled: (1) Exact and quantitative pharmacodynamic CNS tests, allowing of repeated measurements. (2) Highly sensitive and exact methods to assay plasma levels of drugs at repeated intervals (3) Repeated measurements over a prolonged period of time in order to allow ascertaining an exact and detailed time-course of effects and of blood levels over an extended period of time, e.g. necessary to pick up postdrug effects, to enable the determination of rates of absorption and elimination and to pinpoint the height and time of peak effects for tests and for plasma levels and the time, when tests have returned to the initial predrug level. (4) Statistical techniques allowing the evaluation of the significance, reliability and validity of the methods used and of the results achieved.

With this background and with these four criteria in mind we decided to conduct a series of experiments in our laboratories to elucidate the problem of CNS effect – drug level relationships, especially stressing the necessity to follow the events to be studied by repeated tests over a prolonged period of time.

We have earlier used alcohol as a reference standard, a great number of different CNS effects having been studied in relation to blood ethanol (1–5).

For the present series of experiments diazepam was chosen as a model substance, as an example of a psychoactive drug where highly refined and accurate techniques are available for assessing drug plasma levels. For the present series two different CNS test instruments were chosen, one in order to follow possible changes in a basic, physiologically highly integrated system, namely standing steadiness (body posture), by means of statometry (4, 5), and one for studying pursuit skill, probability reasoning, and traffic-oriented psychomotor performance, as observed with a stressalyzer device (1).

The *aims* of the present study were (1) to follow the time-course of diazepam-induced selected CNS functions in relation to plasma levels, and to ascertain whether a relation exists between CNS effects and plasma levels; (2) to compare over time the CNS effects of three different dose levels of diazepam versus placebo, in relation to drug plasma levels; (3) to devise a technique for comparing possible differences in efficacy, reliability and validity of different types of CNS-based tests, as related to their capacity to respond to small changes in drug plasma levels and in dose.

The experiments have been divided into two series, one using statometry (I), the other the stressalyser device (II) for recording CNS effects.

Material, Methods and Procedure

Material. Five (series I) and ten (series II) healthy males, age 18–55 years, volunteered for the experiments. The subjects were checked by medical examination and biochemical screening, had been informed before hand and had freely given consent.

Experimental design. The various doses of diazepam (Hoffmann-Là Roche, Canada), 10 and 20 mg (series I) or 10, 20 and 30 mg (series II), or a placebo, were given double-blind in each series in a random, complete crossover design with a 14-day intervention between treatments.

A zero time urine sample was collected prior to drug administration. The dose was ingested intact with 100 ml of water, the subjects being fasted and remaining ambulatory during the first 3 h of each test day. A light meal was served 4 h after intake.

Blood samples were drawn at frequent intervals during the first 8-hour period following each dose (fig. 1, 2). CNS tests were scheduled to coincide with the blood sampling program. The samples were centrifuged and the separated plasma was analyzed for diazepam by a gas liquid chromatographic technique (10).

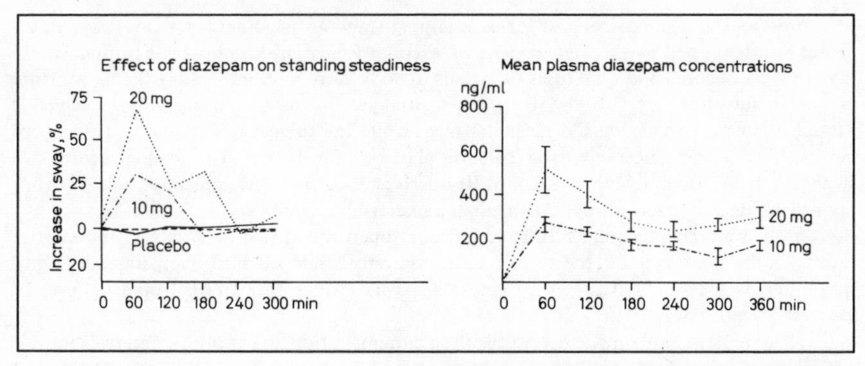

Fig. 1. Time-course of plasma levels and effects of diazepam on standing steadiness (n = 5). Two doses – 10 and 20 mg – and placebo. Plasma levels: means of 5 subjects. (I = ± SD). Standing steadiness: change in body sway (means of sway with open and closed eyes in the lateral and sagittal directions), expressed as percent change of individual, initial predrug levels. Upward deflection: impairment; downward deflection: improvement.

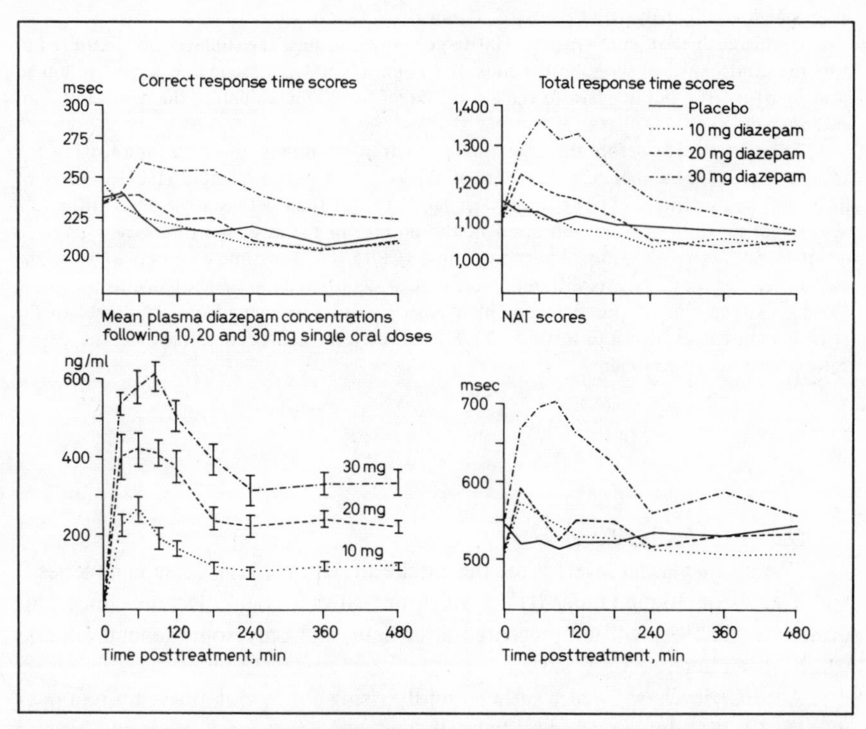

Fig. 2. Time-course of plasma levels and effects of diazepam on parameters of stressalyser changes (n = 10). Plasma levels: means of 10 subjects. (I = ± SD). Stressalyser data: CRT, NAT and TRT. Upward deflection: impairment; downward deflection: improvement.

Statometry. The device used for ascertaining standing steadiness or body sway, Statometer IV, developed at the Department of Alcohol Research, Karolinska Institute, Stockholm (4, 5), records and quantifies variations in body sway by transforming the movements of the center of gravity – both lateral and sagittal sway – into a varying voltage, which is stored on-line on an analog FM magnetic tape, while the subject is standing in a Romberg position on a transducer-equipped balance platform for 70 sec. The analog records are analyzed at the Department of Alcohol Research after A/D conversion; amplitude, variability and frequency spectrum were the main parameters evaluated.

Sway was recorded under four conditions: open and closed eyes, lateral and sagittal sway. In the present work the four conditions were combined into one parameter. Only amplitude is illustrated (fig. 1, 4), expressed as the change from initial, predrug level in percent of initial sway.

Measurements were made before the drug administration and at hourly intervals for 6 h after intake (fig. 1); all subjects had been trained on the instrument before the actual experiment.

Stressalyser device. The Stressalyser, developed in the Control Systems Laboratory, National Research Council (NRC), Ottawa, is capable of detecting subtle differences in decision-making parameters and psychomotor coordination. The device consists of a tracking unit with a lamp display and a control unit. Illumination of one of five lamps, the sequence determined by the control unit, provides the target. Information is stored on magnetic tape and analyzed at the NRC, Ottawa.

During each trial, consisting of 100 target presentations, the subjects were required to grip the control wheel with both hands and align a pointer attached to a reverse linked control wheel; the pointer had to remain on target for 200 msec before the next target was illuminated.

Trials were made before the drug intake and at 30- to 60-min intervals for a total of 8 h after it (fig. 2). A number of parameters of tracking skill were evaluated (1). Here three of them will be presented. (1) The interval or lag (latency) time between the observation of a new target position and the initiation of the pointer in the direction of the new target is denoted as correct reaction or response time (CRT). (2) The time required to steer the pointer to the next target without overshoot is denoted nonovershoot acquisition time (NAT). (3) The time required to complete a successful movement from one target to the next is denoted total response time (TRT). All subjects had been trained on the device before the actual experiment.

Results

Diazepam Plasma Levels

The mean plasma levels after the intake of 10 or 20 mg diazepam (series I) and 10, 20 or 30 mg (series II) are given in figures 1 and 2. Details concerning drug bioavailability will be presented elsewhere, and only some general remarks will be given here.

All dose levels showed a rather rapidly rising absorption phase up to a peak within 30–90 min, an exponential fall, and a subsequent horizontal, or even slightly rising course at 4–6 h after intake, the course still being horizontal up to 8 h.

The results clearly indicate how peak plasma levels as well as the whole course, followed over 6–8 h respectively, are correlated to the dose taken (fig. 1, 2).

Standing Steadiness (Statometry)

Under normal circumstances, with the body erect and standing still, there exists a small reproducible body sway activity, higher in the lateral than in the sagittal direction, and higher with closed eyes than with open eyes (5). The statometer values observed with the placebo dose did not appreciably deviate from the pretreatment baseline values throughout the time of the experiment (fig. 1).

Under the influence of either 10 or 20 mg of diazepam there was a clear impairment in standing steadiness. The time of greatest impairment coincided with the time when peak plasma concentrations of the drug were reached. The course of body sway did not follow the plasma drug levels very closely, but the two doses caused different dose-dependent effects (fig. 1). Standing steadiness returned to baseline values at 3 and 4 h posttreatment respectively for the 10- and 20-mg doses of diazepam.

Stressalyser Data

In the placebo experiments (fig. 2) all parameters (CRT, NAT, TRT) showed an initial improvement, and then a mainly horizontal course, or even some late-occurring impairment (NAT). The time-courses of diazepam plasma levels are given in figure 2, and the time-courses of the effects with the Stressalyser device are illustrated in figures 2–4.

The CRT corresponding to the lag or latency time between observation and response to a new target, showed initially no change after 10 mg of diazepam and after some hours even an improvement as compared to the placebo situation (fig. 4). The 20-mg dose of diazepam produced some prolongation, i.e. impairment. This was pronounced with the 30-mg dose, and leasted for more than 8 h.

NATs, the time required to steer the pointer to the next target, were increased during the first 3 h with the 10- and 20-mg doses. Then the 10-mg dose even produced a marked improvement, as compared to the placebo condition. The 30-mg dose produced very large increases in acquisition times; the impairment was maintained for the duration of the testing period (fig. 2). Essentially the same picture is seen with regard to TRT (fig. 2–4): a nonsignificant impairment, and a subsequent improving effect after 10 mg, an impairment after 20 mg, and a significant, long-lasting impairment after 30 mg of diazepam.

From the results obtained it appears that, under the influence of diazepam and related to the dose and to the change in existing plasma levels, subjects took longer to react to stimuli (new target presentation) and moved more slowly during the acquisition of the new target. Error rates (turning the pointer in the

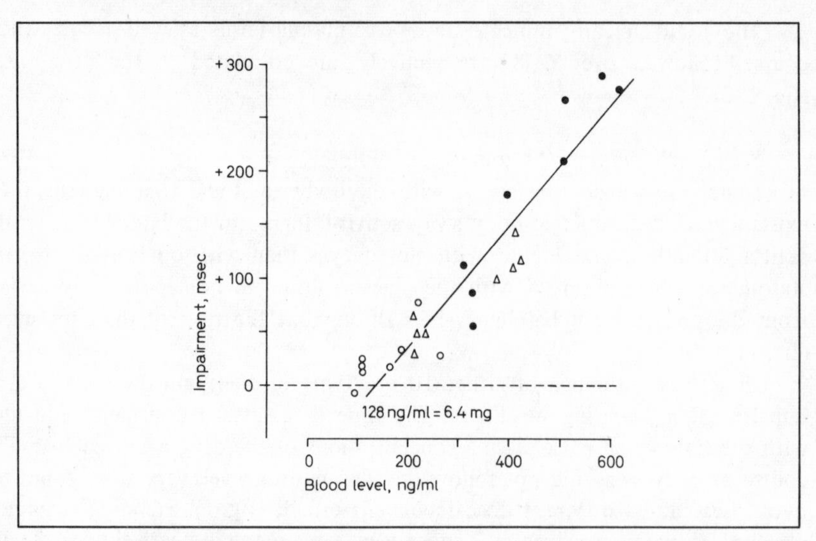

Fig. 3. Relation between TRT scores and diazepam plasma levels. TRT scores are defined as the differences between actual TRT times recorded and the initial, predrug TRT times, corrected for changes in the placebo situation, i.e. diurnal changes. Each point represents the means of TRT scores, plotted against the corresponding mean plasma level found for each dose level tested. Values above 0-line indicate impairment, and below 0-line improvement. Rectilinear regression line based on method of least squares, based on 240 TRT scores and 240 plasma levels. Regression and correlation parameters given in table I. ○ = 10 mg; △ = 20 mg; ● = 30 mg.

wrong direction, i.e. the direction away from the new target), however, were not increased with diazepam. These results suggest that while motor function is impaired with diazepam, correct decision generation is unaffected at the doses studied.

Assessment of a Possible Relation between CNS Effects and Drug Plasma Levels

Due to the time-course being followed by quantitative methods over an extended period of time by repeated determinations of CNS effects as well as of drug plasma levels the possible relation between test results and plasma levels could be studied by means of an extensive regression and correlation analysis. The analysis was originally based on all the individual values obtained – 360 in series I and 960 in series II, or a total of 1,320 test values and 410 plasma values; in the present study only the means are illustrated (fig. 2–4).

The test data – statometer data in percentual change from initial, predrug performance, and CRT, NAT and TRT in milliseconds – were corrected for differences in initial, predrug performance and for diurnal variations in the place-

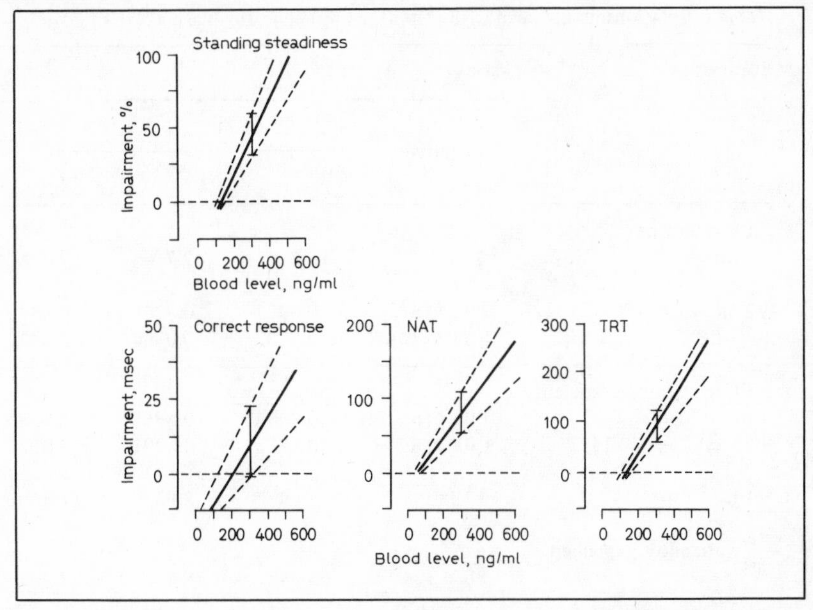

Fig. 4. Relation between standing steadiness (body sway) or stressalyser scores (CRT, NAT, TRT) and corresponding diazepam plasma levels. Regression lines based on method of least squares, regression, correlation and threshold parameters found in table I. Errors of estimate (= $\pm s_{yx}$) are marked on the graph, as well as threshold values (=x-intercepts). Regression lines based on 240 determinations at three dose levels, 10, 20 and 30 mg, 80 determinations per dose level. Values above threshold (= 0-line) indicate impairment, below threshold improvement. Note that the four tests have different threshold values, different slopes and different error estimates (table I). CRT is improved on diazepam blood levels below the threshold, corresponding to an intake between 5 and 10 mg diazepam; also standing steadiness and TRT show a tendency to improvement at blood levels below their thresholds.

bo situation (fig. 2) by using the *differences* (stressalyser data) between a value obtained at any postdrug determination and the initial, predrug value, corrected for the possible change in the placebo performance at that time.

The first stage of this analysis is illustrated in figure 3, where means of TRT scores, i.e. the differences between actual TRT times and the initial, predrug times, corrected for variations in the placebo performance (fig. 2) are plotted against the mean drug plasma levels at corresponding times, each dose denoted by a different mark (fig. 3).

The rectilinear regression line, calculated by the method of least squares, is given in figure 3, as well as its x-intercept. No significant differences in slope or position between the regression lines for the three dose levels were found in this material; hence the final regression analysis was based on the values for all the

Table I. Relationship between CNS test effects (y) and drug plasma levels (x)[1]

Parameters	Tests (n = 24)			
	statometry % of initial	stressalyser, msec		
		CRT	NAT	TRT
1 y-Intercept (a)	−31.7	−19.8	−17.7	−69.4
2 Slope (b) ± SE	0.2538 ± 0.0194	0.0972 ± 0.0152	0.3249 ± 0.0364	0.5392 ± 0.0431
3 Correlation coefficient r $\pm e_r$	0.9416 ± 0.0718	0.8065 ± 0.1261	0.8853 ± 0.0991	0.9363 ± 0.0721
4 t_r	13.11	6.40	8.93	12.50
5 Variability explained, % r^2	88.7	65.0	78.4	87.7
6 Error estimate s_{yx}	± 12.44	± 11.08	± 26.51	± 31.42
7 Minimal significant change in effect (y) $2.064 \times s_{yx}/\sqrt{24}$	+ 4.40	+ 4.43	+ 10.61	+ 12.57
8 Corresponding change in blood level (x), ng/ml	+ 17.32	+ 45.55	+ 32.66	+ 23.31
9 Corresponding change in dose mg drug/70 kg ± SE	+ 0.87 ± 0.13	+ 2.28 ± 0.33	+ 1.63 ± 0.24	+ 1.17 ± 0.17
10 Critical level (threshold) = x-intercept; ng/ml ± SE (mg drug/70 kg)	124.8 ± 15.5 (6.2 mg)	204 ± 27.9 (10.2 mg)	54.6 ± 32.4 (2.7 mg)	128.7 ± 18.3 (6.4 mg)

[1] Rectilinear regression, method of least squares. Formula: y = a + bx, where a = y-intercept, and b = slope.

three dose levels tested, comprising a total of 24 mean values for each parameter in the stressalyser series, the analysis here being based on means.

Slopes (b), y-intercepts (a), correlation coefficients (r) and error estimates (departure from linear regression line s_{yx}) were computed, including corresponding t values (table I). A highly significant correlation (table I, line 3) between test data and plasma drug levels was seen for all parameters tested, all slopes being positive and highly significant (line 4), indicating a highly significant increase in degree of impairment with increase in plasma drug level ($p < 0.001$). The change in drug level thus explains between 65 and 88.7 % of the total variability (line 5), depending on the parameter tested.

Comparison between Different CNS Test Parameters

In order to compare the efficiency of different CNS test parameters the regression lines for all tests separately were calculated. They are illustrated in figure 4, together with the error estimate (s_{yx}); all relevant statistical data are found in table I.

A comparison between the different parameters cannot be based on the y-intercepts (line 1), the slopes (line 2), or the error estimates (line 6), as they stand, because these parameters are measured in different units (statometer versus stressalyser), or in units of different order of magnitude (CRT, NAT, TRT; fig. 2).

A comparison of the correlation coefficients, which are not expressed in any measures, their errors and t values (lines 3–4), however, show clear variations between the test parameters, statometry showing the highest and CRT the lowest values; the same rank order is also seen from the fraction of variability explained (line 5).

In order to make a more intricate comparison possible, and at the same time to express the possible differences between the tests in some common unit, the following technique was employed. As the separate regression lines were based on 24 pairs of values, the minimal significant change in degree of impairment was calculated as:

$$t_{p = 0.05} \times s_{yx} / \sqrt{24}, \text{ d.f.} = 22.$$

The resulting values, expressed in respective units are given in table I (line 7).

The corresponding changes in the drug plasma level – expressed in ng/ml, hence the same unit for all tests – were calculated from the respective regression lines (table I, line 8). The average doses of diazepam, necessary to bring about the minimal, significant changes in plasma levels, were also calculated (line 9), in mg/70 kg body weight, based on 10 mg diazepam per 70 kg giving average plasma levels of 200 ng/ml (fig. 1, 2).

A highly significant, differential order of magnitude ($p < 0.001$) between the four parameters tested is now observed. Statometry and TRT show the smallest variability, the highest correlation and the smallest 'minimal' significant change in blood level or dose, whereas CRT shows the greatest variability. This technique thus allows comparison among different types of tests, and shows a greater differentiation than, using the correlation coefficient, for example.

Critical Level (x-Intercept, Threshold)

While the slope (b) of a dose-response curve is an indication of the rate at which effects change with changes in plasma levels, the point where the regression line crosses the 0-level – the x-intercept – is an expression of the critical level (2, 4), or threshold – expressed in ng drug/ml – where an impairment, as measured by that specific test, begins (appearance threshold), or terminates (disappearance threshold). The critical levels and their errors were calculated from the regression equation and expressed in ng/ml (line 10), and in milligram drug taken per 70 kg (line 11) (fig. 4).

All test parameters used showed a high degree of sensitivity, i.e. being able to detect changes in performance at a minimal level of 54.6–204 ng/ml, or 2.7–10.2 mg diazepam/70 kg. NAT was the most sensitive and CRT the least sensitive in this respect, statometry and TRT being in between (lines 10, 11).

It is also interesting to note that the part of the regression line (fig. 4) that goes below the abscissa indicates that diazepam in doses *below* the threshold value, e.g. for CRT below 204 ng/ml, or less than 10.2 mg/70 kg, has a stimulating or improving effect, changing into a depressant impairing effect at doses over the threshold. The same is also suggested for statometry and TRT, at doses below 5 mg. The same phenomenon – an initial stimulating, or improving effect – was also observed in the actual time-course curves (fig. 2), when these were compared to placebo performance.

Summary

Pharmacodynamic effects and plasma levels of diazepam were studied in healthy male volunteers at different dose levels. Responses to diazepam were quantified, using instruments which measured body sway (statometry) and psychomotor performance (stressalyser tests). High dose-related correlations were obtained between drug-induced changes in test parameters and drug plasma levels, both with regard to stimulant and depressive effects. Techniques were devised for evaluating and comparing the efficacy and usefulness of different types of tests, taking into account critical thresholds, slopes and error estimates, correcting for changes in predrug levels and control (nondrug) trials.

Comment by Dr. M.J. Mattila

Thank you, Dr. *Goldberg*, for this presentation which combines both pharmacodynamics and pharmacokinetics in the same study. However, I have some comments about the interpretation of the results. Your statistical analysis of the correlation of psychomotor skills and diazepam plasma levels may well apply to late effects of diazepam after acute administration, as well as to long-term treatment. But the analysis may not apply, say, to the first 2 h after the diazepam intake when the pharmacodynamics do not correlate with the pharmacokinetics. This is well seen in figure 1 where the impairment in body sway after the large dose of diazepam has subsided to the same as, or less than, the performance after the lower dose of diazepam, yet the plasma levels of diazepam was definitely higher after the large dose. This discrepancy well agrees with the observations of our group. It may be that a better agreement between pharmacodynamics and pharmacokinetics, even after a single dose, is obtained during long-term treatment where the receptor tolerance probably damps the initial response during the first hour after the single dose. But we do not know if this really happens, and we should design trials to elucidate the problem.

References

1 *Dussault, P. and Chappel, C.I.:* Differences in blood alcohol levels following consumption of whisky and beer in man. Proc. 6th Int. Conf. Alc. Rd. Traff., Toronto 1974.

2 *Goldberg, L.:* Quantitative studies on alcohol tolerance in man. Acta physiol. scand. *5:* suppl. 16, pp. 1–128 (1943).

3 *Goldberg, L.:* Alcohol tranquilizers and hangover. Q. Jl. Stud. Alcohol *1:* 37–56 (1961).

4 *Goldberg, L.:* Behavioral and physiological effects of alcohol on man. Psychosom. Med. *28:* 295–309 (1966).

5 *Goldberg, L.:* The interaction of alcohol and other CNS-acting drugs in man and animals; in *Saffron* Alcoholism, vol. 1, pp. 6–46 (Academy of Medicine, New Jersey 1972).

6 *Haffner, J.F.W.; Morland, J.; Setekleiv, J.; Strømsaether, C.E.; Danielsen, A.; Frivik, P.I., and Dybing, F.:* Mental and psychomotor effects of diazepam and ethanol. Acta pharmac. tox. *32:* 161–178 (1973).

7 *Korttila, K. and Linnoila, M.:* Skills related to driving after intravenous diazepam, flunitrazepam or droperidol. Br. J. Anaesth. *46:* 961–969 (1974).

8 *Korttila, K. and Linnoila, M.:* Psychomotor skills related to driving after intramuscular administration of diazepam and meperidine. Anaesthesiology *42:* 685–691 (1975).

9 *Linnoila, M. and Mattila, M.J.:* Drug interaction on psychomotor skills related to driving: diazepam and alcohol. Eur. J. clin. Pharmacol. *6:* 186–194 (1972).

10 *Da Silva, J.A.F. and Puglisi, C.V.:* Determination of medazepam (Nobrium), diazepam (Valium) and their major biotransformation products in blood and urine by electron capture gas liquid chromatography. Analyt. Chem. *42:* 1725 (1970).

11 *Wallgren, H. and Barry, J., III:* Actions of alcohol, vol. 1, 2, pp. 871 (Elsevier, Amsterdam 1970).

Prof. *J. Orr*, Bio-Research Laboratories, *Pointe-Claire, Que.* (Canada)

Mod. Probl. Pharmacopsych., vol. 11, pp. 68–78 (Karger, Basel 1976)

Precision, Accuracy and Relevance of Breath Alcohol Measurements[1]

A. W. Jones

Department of Alcohol Research, Karolinska Institutet, Stockholm

A physiological relationship exists between the concentration of alcohol in end expired alveolar air and in the blood. This is defined by the *in vivo* blood/breath partition ratio which is generally considered to be 2,100 to 1 at an expired breath temperature of 34 °C (2). As a biological parameter the blood/breath partition ratio is subject to interindividual and intraindividual variations (9). Nevertheless, during the post-absorption phase of ethanol metabolism the variability is sufficiently low for breath alcohol analysis to be used as an estimator of the blood alcohol concentration (BAC).

A wide range of methods are available for breath alcohol analysis and depending on the analytical principles involved either a qualitative or a quantitative measure of the underlying blood alcohol concentration is possible. The current interest in breath alcohol analysis as a routine procedure in clinical and medicolegal work has stimulated detailed studies into reliability and usefulness of breath-testing equipment for this purpose.

For many years elaborate programs have been used in our laboratory for evaluating breath-testing instruments, comprising the following experimental stages. (1) *In vitro* studies using breath simulator techniques under various conditions with air-alcohol mixtures of known concentration. (2) *In vivo* experiments with subjects given known doses of alcohol in a laboratory environment. (3) Field studies under double-blind conditions with laboratory control over blood sampling and analysis. (4) Roadside testing of motorists.

This communication concerns recent findings with regard to the physiological background and the experimental and statistical techniques suitable for evaluating the reliability of a wide range of new breath-testing devices. Detailed reports of each individual instrument will be published elsewhere.

[1] These studies have been defrayed by grants to Prof. *L. Goldberg* from the National Police Board (Rikspolisstyrelsen) and the Swedish Medical Research Council (grant 552).

Table I. Breath alcohol instruments evaluated

Instrument or testing device	Operating principle
Mark II gas chromatograph intoximeter	gas-liquid chromatography using a flame ionisation detector
Alcohol screening device (ASD)	fuel cell detector
Breathalyzer model 900	photometric measurement of potassium dichromate reduction
Alcolmeter pocket instrument	fuel cell detector
Alco-Limiter	electrochemical oxidation
Intoxilyzer	infra-red absorption
Alcotest 0.5 and 0.8 ‰ ampoules	visible colour change after potassium dichromate reduction
Alcolyser 0.5 and 0.8 ‰ ampoules	visible colour change after potassium dichromate reduction

Methods

Instruments for Breath Alcohol Analysis

The breath-testing instruments evaluated within this program are shown in table I, as well as the analytical principles used for the detection and analysis of ethanol in breath.

Alcohol Vapour Standards

The errors inherent in the analytical methods employed in each breath instrument have been investigated using air-alcohol vapour standards. Such standards are also used for instrument calibration before testing subjects. Comparisons have been made between a dynamic breath simulator method operating at 34 °C and a new static method based on sampling the equilibrium head space vapour above a known strength aqueous ethanol solution (8).

In vivo *Studies*

Healthy male subjects (n = 55) have been used in the present series of experiments. In general, whisky 36 % w/v was used thoughout the present series and given in a standard dose of 2 ml/kg body weight equal to 0.72 g alcohol/kg. The calculated dose was required to be consumed neat over a period of 20 min. In order to 'standardise' alcohol absorption kinetics the subjects were required to be in a fasted condition.

Blood sampling. A blood sample was taken immediately before drinking to check alcohol-free status and thereafter capillary blood samples were taken in triplicate at regular intervals throughout the whole blood alcohol course. The sample volume (10 μl) was measured using disposable blood pipettes and transferred into Autoanalyser cups containing 1 ml of 0.05 *M* sodium fluoride.

Breath sampling. Breath samples were taken in accordance with the requirements for each specific instrument with emphasis on obtaining end expired alveolar air. The standard procedure was to take one breath sample before and one after the blood sampling, the time difference between the two breath samples being only 2 min. Samples were taken before drinking, at 30-min intervals for the first 2 h after drinking, and thereafter at 1-hour intervals until the subjects were alcohol-free, usually for a total of 7–9 h. This procedure enables each phase of alcohol metabolism to be carefully monitored, and in particular to monitor the instrument performance when the subjects were around the critical legal limit, 0.5, 0.8 or 1 mg/ml (‰), respectively, and when they were approaching zero blood alcohol.

Blood alcohol analysis. The alcohol concentrations from blood samples were measured using an automated enzymatic micro-technique (4), recently modified to include a distillation step for increased sensitivity (3). This technique has the capacity to analyse 50–60 samples per hour enabling 400–500 independent blood samples to be run each day. The analytical precision for triplicate samples of blood has been shown to be 0.01 mg/ml at a mean concentration of 0.5 mg/ml (3) – 0.5 mg/ml (‰) is equivalent to 50.0 mg/100 ml.

Results

In vitro *Studies*

In vitro experiments are the logical first step in evaluating a breath alcohol instrument since biological variation is eliminated. The merits of the analytical principle and the precision of the sampling mechanism can be investigated under these circumstances. Precision can be estimated from multiple analysis at fixed concentration levels and linearity determined over relevant concentration ranges. To illustrate an example from this stage of testing, some results with the Gas Chromatograph Intoximeter are reported. Figure 1 shows the instrument response measured in peak area units in relation to the blood alcohol equivalent concentration of the two alcohol vapour standards. Ten measurements at each concentration level for both the simulator and head-space standards were made.

The relationship between detector response and concentration for both the simulator and head-space methods of preparing air-alcohol standards was first determined by fitting individual least squares regression lines for each standard. As shown by covariance analysis the slopes as well as the elevations of these two lines were not statistically different, $p > 0.05$ in each case. This implies that they are samples from the same parent population enabling a pooling of the results and the fitting of a single rectilinear regression line. The regression equation was $y = 0.433 + 35.8\ x$ and the y-intercept 0.433 was not significantly different from zero, $t = 0.522$ (d.f. = 6, $p > 0.05$).

A measure of the *in vitro* precision of the gas chromatograph intoximeter is furnished by the variability of the points around the least squares fitted line. In statistical terms this is given by s_{yx}, the error estimate, being ± 0.977 peak area units or 3.70 % of the mean. Dividing s_{yx} by the regression coefficient b gives the variability of the corresponding x-variate (1) being ± 0.027 mg/ml or 3.72 % of the mean blood alcohol equivalent concentration.

In vivo *Studies*

Precision of breath alcohol analysis. As an example of how the *in vivo* precision of breath alcohol instruments has been determined, some results from experiments using the alcohol screening device (ASD) will be presented.

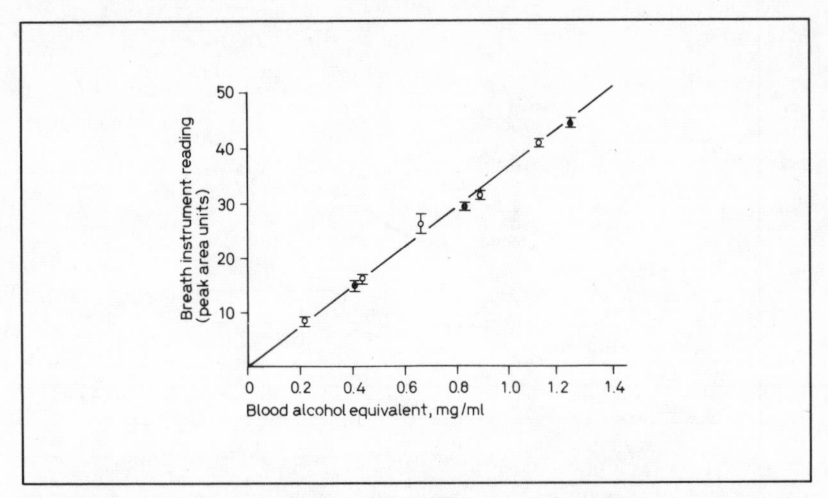

Fig. 1. The *in vitro* relationship between peak area and alcohol vapour concentration using a Gas Chromatograph Intoximeter and air-alcohol standards prepared by two independent methods. The mean and range of 10 determinations at each concentration level are plotted. • = Simulator; ○ = head space.

The short interval of 2 min between breath sampling means that each pair of readings may be considered as duplicates, the differences between these duplicates were used to estimate the precision and experimental errors of the method. A total of 147 duplicate breath samples were analysed, ranging from 0.04 to 1.30 mg/ml (mean 0.50 mg/ml). The variability (SD) of a single determination was calculated from the SD of the differences between duplicate results according to the following formula:

$$s_x = \frac{\sqrt{\frac{S(d - \bar{d})^2}{n - 1}}}{\sqrt{2}}, = \frac{Sd}{\sqrt{2}}$$

where d = difference between duplicate determinations; $\bar{d}$ = mean of the differences; n = number of differences; s_d = SD (variability) of the differences; s_x = SD (variability) of a single determination.

This procedure requires that the individual differences are randomly distributed with a mean not significantly different from zero. This is proved statistically using a Student's t test, the mean difference was -0.0031 ± 0.0064 mg/ml; $t = -0.4830$ (d.f. = 146, $p > 0.05$) and confirms the random nature of these differences. The SD of the differences was found to be ± 0.0753 mg/ml and the precision of a single determination ± 0.0532 mg/ml.

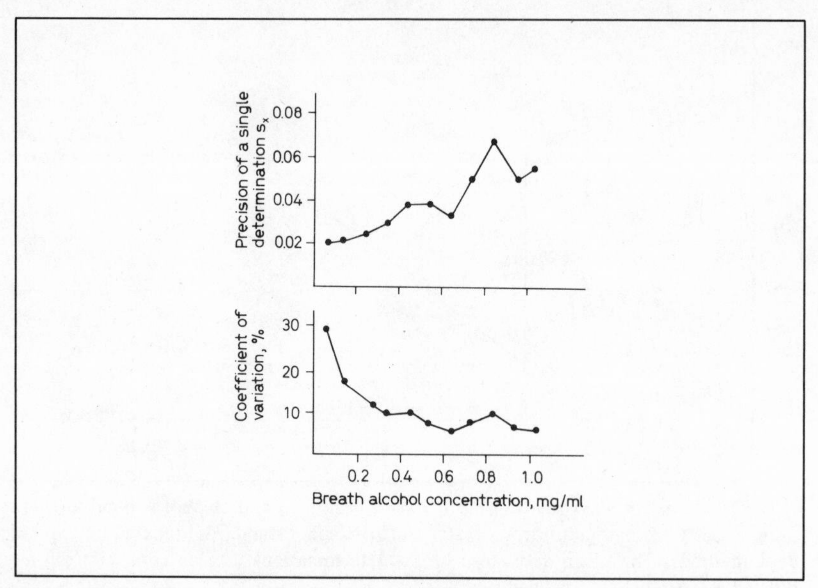

Fig. 2. The precision of breath alcohol analysis as a function of concentration using the ASD. Results based on duplicate determinations.

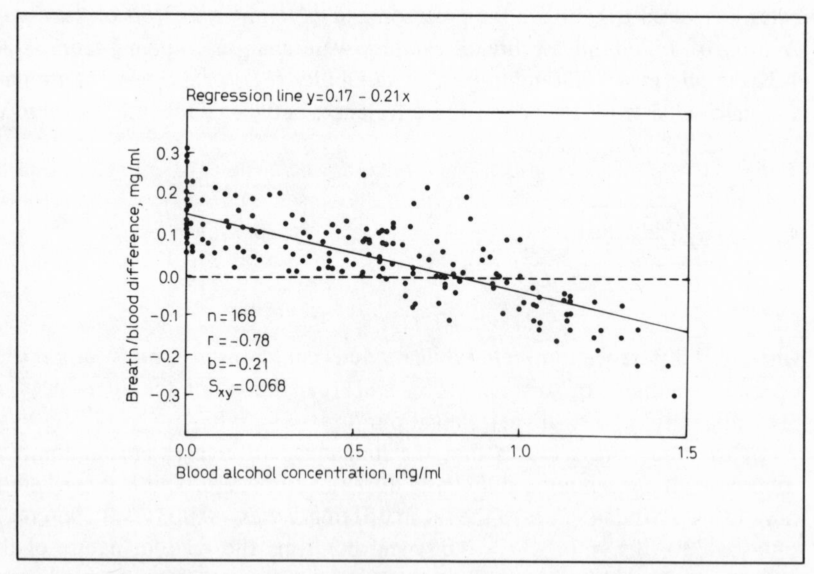

Fig. 3. The breath/blood alcohol concentration difference in relation to the blood alcohol concentration using the ASD for breath analysis.

Since in this work all analyses were based on duplicate determinations, the experimental error is defined as the SD of a double determination, and is calculated as:

$$\frac{0.0532}{\sqrt{2}} = \pm\ 0.038 \text{ mg/ml.}$$

In order to see whether the precision of breath analysis is related to the breath alcohol concentration, precision was related to breath alcohol over a range of concentrations. Figure 2 shows the variability of a single determination in relation to concentration in both absolute (upper part) and in relative terms (lower part).

The precision varies essentially inversely with the variability. The precision in relative terms, i.e. the coefficient of variation, of a single determination was found to be 10.6 % at a mean concentration of 0.5 mg/ml (fig. 2, lower part).

Accuracy of breath alcohol analysis. As a measure of the accuracy of an analytical technique, it is necessary to consider the nature of the differences between the concentration observed and the true values, whether systematic or random. In these experiments, if the results of blood analysis are taken as the standard, the differences between the blood and breath results serve as a measure of the accuracy of the breath instrument. From a statistical point of view, an accurate technique will be characterised by a random distribution of the differences around a mean of zero.

This statistical technique has been used to analyse the results from a field study using the ASD. The breath/blood differences have been plotted against blood alcohol concentration and are shown in figure 3.

The regression of breath/blood difference on BAC is defined by the equation $y = 0.166 - 0.208\,x$ indicating a negative regression with a highly significant correlation $r = -0.776$ ($p < 0.001$). This particular ASD instrument reads too high at BACs below 0.5 mg/ml and too low above 1.0 mg/ml.

The mean breath/blood difference was found to be + 0.051 mg/ml but the individual differences show a *trend* when one considers them in relation to BAC. For example, at zero BAC, the ASD gave readings significantly greater than zero. The maximum ASD difference was found to be + 0.32 mg/ml when the true blood alcohol concentration was zero. The mean zero reading, i.e. the regression equation y-intercept, was 0.166 ± 0.009 mg/ml, $t = 18.44$ ($p < 0.001$).

Blood/breath correlation. The precision and accuracy of breath alcohol instruments may be analysed further using a blood/breath scatter diagram. The results of extensive experiments using the Breathalyzer (5) are shown in figure 4.

The appropriate technique for a mathematical analysis of these data is that of regression analysis, using the BAC as the independent variable. In figure 4, the regression equation was found to be $y = 0.012 + 1.02\,x$ and breath and blood

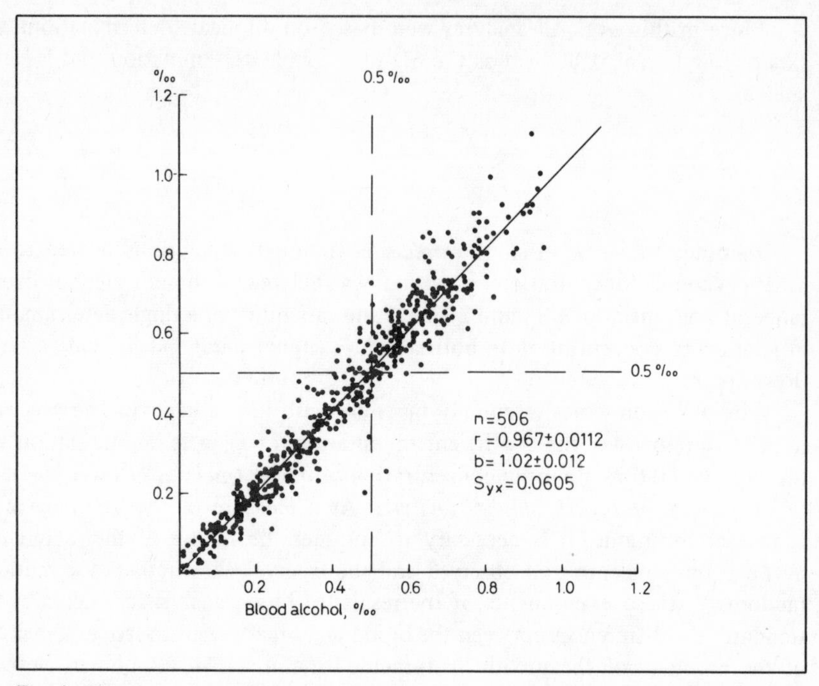

Fig. 4. The regression relationship between simultaneous blood and breath analysis using a Breathalyzer 900.

were highly correlated $r = 0.967 \pm 0.011$ ($p < 0.001$). The regression coefficient 1.021 ± 0.012 ($p < 0.001$) was not significantly different from unity and the y-intercept 0.012 not significantly different from zero, i.e. a perfect 1:1 relationship in this study, there being no systematic differences between the breath and blood alcohol concentrations within the range of BAC levels used. The variability of the individual points around the least squares regression line s_{yx} was low, ± 0.0605 mg/ml, and is a measure of the high accuracy of this instrument.

From this figure of ± 0.0605 the 95 % confidence limits for unbiased estimates of the blood alcohol concentration using this instrument may be shown to be ± 0.13 mg/ml at a mean breath alcohol result of 0.50 mg/ml. This means that if a random breath sample is taken from another subject, there is only a 1 in 20 chance that it will differ by more than ± 0.13 mg/ml from the true blood concentration at the 0.5 mg/ml level.

The blood/breath ratio in relationship to the phase of alcohol metabolism. The design of the present experiments involved serial blood and breath alcohol measurements for the entire time alcohol was in the body, i.e. during the absorp-

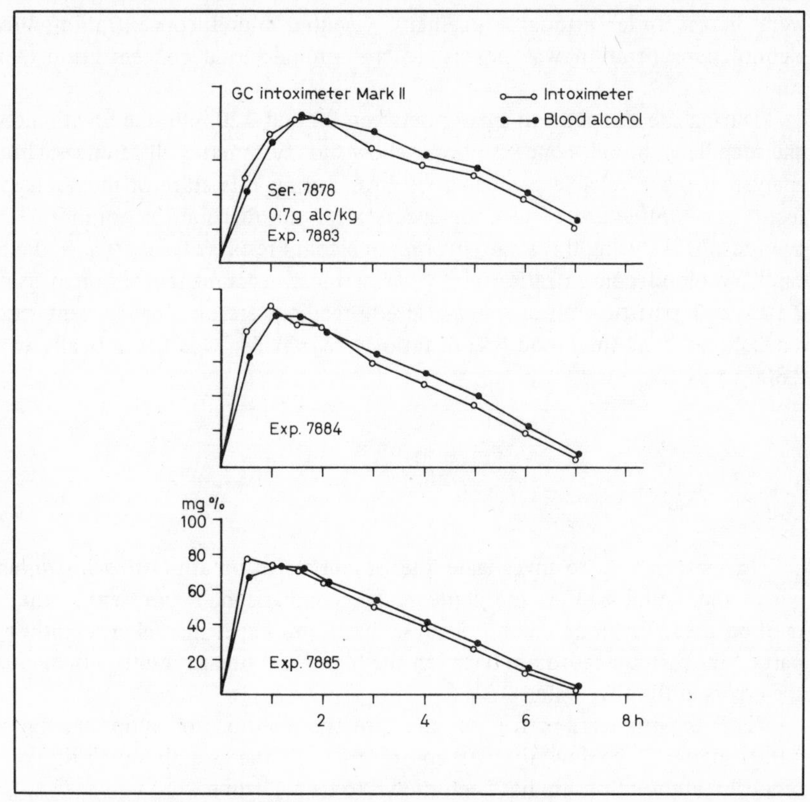

Fig. 5. The alcohol concentration time-course for both breath and blood samples during the absorption, distribution and elimination phase of metabolism in three subjects. Tests were made on the same day and with the same instrument calibrated using a 2,100 to 1 blood/breath ratio.

tion, distribution and elimination phases of metabolism. This enables a study of the variability of the *in vivo* blood/breath relationship. Some typical curves for three different subjects are shown in figure 5. The Gas Chromatograph Intoximeter was used for breath analysis and calibrated using a 2,100 to 1 blood/breath partition ratio.

During active *absorption* of alcohol, in these experiments occurring between 30 and 90 min from the start of drinking, the BAC predicted by breath analysis was somewhat *higher* than that found by capillary blood analysis. The breath alcohol concentration reflects pulmonary arterial blood concentration and during conditions of rapid absorption there is an arterio-venous alcohol difference. *Harger et al.* (6) showed that during the absorption phase, the alcohol concentration in samples of blood from different parts of the vascular system were in the order arterial > capillary > venous blood concentration; breath alcohol concentration was nearest to the arterial blood concentration in this study.

During the *distribution* phase, between 90 and 120 min, the breath alcohol and capillary blood concentrations show *no* systematic differences, breath analysis being a reliable estimator of BAC during this stage of metabolism. A rectilinear *elimination* of alcohol occurs after the distribution equilibrium has been reached. During this stage, the breath alcohol results *underestimate* the true capillary blood concentration (fig. 5). For most instruments calibrated using a 2,100 to 1 partition ratio systematic *low* readings are obtained; recent studies have shown that the blood/breath ratio *in vivo* at 34 °C is theoretically nearer 2,300 to 1 (8).

Discussion

In experiments to investigate the properties of breath-testing instruments, where the actual BAC is the standard for comparison, it is essential that the method used for blood alcohol analysis has a low experimental error, otherwise part of the differences found between the blood and breath results will be due to the errors in the BAC values.

The experimental error of the present method of blood analysis of ± 0.01 mg/ml is so low that its contribution to the overall variability of the blood/breath relationship has been shown to be negligible.

If at a mean blood alcohol concentration of 0.5 mg/ml the precision in relative terms is taken as ± 2 %, i.e. precision as a percentage of the mean, then from the blood/breath correlation diagram (fig. 4) with an error estimate of ± 0.065 mg/ml or 13 % of the mean concentration, the variability unexplained and resulting from biological sources is given by:

$$\sqrt{13.0^2 - 2.0^2} = \pm 12.8\ \%,$$

i.e., the error in blood analysis has an insignificant contribution.

The results from *in vitro* experiments will reflect the maximum limits of precision and accuracy possible for each particular instrument since there is no human variables to be considered. The reliability of the methods of preparing alcohol vapour standards has been confirmed in these experiments by using two independent physical principles, i.e. static and dynamic equilibration operating at two different temperatures, i.e. the vapour pressure-temperature relationship for alcohol reported in the literature may be considered correct (7, 8).

The technique of simultaneous blood and breath analysis and serial sampling from individual subjects enables an exact mathematical expression for the relationship between these two variables to be computed. In addition, the blood/breath relationship can be analysed during each phase of alcohol metabolism and the intra- and inter-subject variability can be investigated.

Our observations imply that the *in vivo* blood/breath partition ratio varies in a regular and systematic way depending on the phase of alcohol kinetics. Theoretically, for a precise estimate of the underlying capillary blood alcohol concentration a different ratio would apply according to the phase of metabolism. If a partition ratio higher than 2,100 to 1 is used, e.g. 2,300 to 1, then a reliable instrument would show no differences during the elimination phase but accentuate the differences during the absorption phase. Conversely, an instrument which happens to show no differences during the absorption phase will show large differences during the elimination phase.

The results found with the ASD are typical of an instrument having a high precision but a low accuracy. The difference between these two terms when applied to an analytical method or breath alcohol instrument need to be clearly defined and differentiated. The precision was ± 0.038 mg/ml for duplicate samples being high considering the many variable factors involved, e.g. breath temperature, breathing pattern of subjects, etc. The accuracy was low, the instrument giving positive readings when there was no alcohol in the blood, and the magnitude and direction of the breath/blood differences changing with the blood alcohol concentration.

The findings from comprehensive evaluation programs, such as that outlined here, enable specifications to be set regarding each particular instrument, covering both *in vitro* and *in vivo* reliability and the practical usefulness of the method. Although there is an inherent biological variability to be considered in any method of breath analysis, the fact that many samples can be taken over a short period of time to give an on-the-spot objective diagnosis of intoxication make a method acceptable for medico-legal work provided that both precision and accuracy are documented and are within the limits of the specifications laid down for each type of breath instrument.

Summary

The methods employed for evaluating the performance of some currently available breath alcohol devices are described. A program consisting of *in vitro* experiments, extensive *in vivo* laboratory studies, controlled field studies, and finally roadside random testing has been used. The importance of considering the underlying physiological mechanisms particularly with regard to the blood/breath partition ratio and its variability is emphasized. An important finding is that the relationship varies with the phase of alcohol metabolism in a regular and systematic way. Some statistical techniques are outlined for calculating the precision and accuracy of breath-testing instruments, which are important properties of a method to be used for medico-legal purposes.

References

1 *Bliss, C.I.:* Statistics in biology, vol. 1 (McGraw-Hill, Maidenhead 1967).

2 *Borkenstein, R.F.; Darwick, N.; Forney, R.B.; Force, R.; Harger, R.N.; O'Neill, B.; Buck, R.C.; Dubowski, K.; Forrester, G.C.; Goldberg, L.; Wright, B.M., and Prouty, R.W.:* Statement; in Proc. of the ad hoc Committee on the Alcohol Blood/Breath Ratio, Indiana University, Ind. 1972.

3 *Buijten, J.C.:* An automated ultra micro distillation technique for determination of ethanol in blood and urine. Blutalkohol *12:* 393 (1975).

4 *Goldberg, L. and Rydberg, U.:* Automated enzymatic microdetermination of ethanol in blood and urine. Proc. Technicon 5th Int. Symp. on Automation in Analytical Chemistry, London 1965.

5 *Goldberg, L. and Bonnichsen, R.:* Bestämning av noggrannheten i alkotestmetoden och vissa andra utandningsmetoder. Bilaga 5. Trafiknykterhetsbrott, Statens Offentliga Utredningar (Sweden) 1970, p. 424 (1970).

6 *Harger, R.N.; Raney, B.B.; Bridwell, F.G., and Kitchel, M.F.:* The partition ratio of alcohol between air and water and urine and blood: Estimation and identification of alcohol in these liquids from analysis of air equilibrated with them. J. biol. Chem. *183:* 197 (1950).

7 *Harger, R.N.:* Blood source and alcohol level, errors from using venous blood during active absorption. Proc. 3rd Int. Conf. Alc. and Road Traffic, London 1962.

8 *Jones, A.W.:* Equilibrium partition studies of alcohol in biological fluids; thesis, University of Wales (1974).

9 *Jones, A.W.:* The variability of the blood-breath partition ratio *in vivo* (in preparation, 1975).

10 *Wright, B.M.; Jones, T.P., and Jones, A.W.:* Breath alcohol analysis and the blood-breath partition ratio. Med. Sci. Law *15:* 203 (1975).

Dr. *A.W. Jones,* Department of Alcohol Research, Karolinska Institutet, *Stockholm* (Sweden)

Mod. Probl. Pharmacopsych., vol. 11, pp. 79–84 (Karger, Basel 1976)

Effect of Active Metabolites of Chlordiazepoxide and Diazepam, Alone or in Combination with Alcohol, on Psychomotor Skills Related to Driving

E.S. Palva, M. Linnoila and M.J. Mattila

Department of Pharmacology, University of Helsinki, Helsinki

Several benzodiazepines with similar pharmacological properties are widely used in practice, and some of them and their metabolites accumulate in tissues during the treatment. Many of the metabolites are pharmacologically active and they have a long half-life (1). In the present study oxazepam, methyloxazepam, *N*-desmethyldiazepam and chlordiazepoxide lactam were investigated concerning their actions and interactions on psychomotor skills related to driving. Psychomotor performance was measured by using methods previously used in this laboratory for the studies on the parent compounds diazepam and chlordiazepoxide (8, 9).

Material and Methods

40 healthy students, 17 males and 23 females, aged 20–29 years, volunteered for a subacute double-blind crossover experiment. Two female subjects had to interrupt the trial and their results were excluded.

Trial design. The subjects were divided into two groups of 20 persons each. One group received oxazepam (O) 15 mg methyloxazepam (MO) 20 mg, and placebo t.i.d. double-blind crossover in random order for 2 weeks each. The other group received similarly *N*-desmethyldiazepam (DMD) 5 mg, chlordiazepoxide lactam (ChL) 10 mg, and placebo, each t.i.d. The drugs were given in identical gelatine capsules. In the beginning of each test session, 30 min before the first test time, the subjects received alcohol 0.5 g/kg (A) or placebo drink together with their drugs.

The psychomotor test battery included a choice reaction test, two coordination tests, an attention test (6, 7), and two tests measuring hand and foot proprioception (4). Flicker fusion and horizontal nystagmus were measured as well. At every test time the subjects were asked to assess their feeling of performance and the nature of their treatment. To avoid the effect of learning, the subjects were trained on the apparatuses for 3 h before the beginning of the treatment. The tests were carried out after taking the first capsule (acute test), and after 1 and 2 weeks' treatment (subacute tests), each time 30, 90, and 150 min after the administration of the drug and drink. One group was tested in the morning beginning at 9 a.m., and the other in the afternoon beginning at 4 p.m.

Other tests. Blood samples were collected on the 1st, 7th and 14th day of each treatment, 3 h after the administration of the drug. The sera were kept at – 20 °C until assayed for drugs by EC-gas chromatography and by spectrofluorometry (3, 10, 11). Student's t test and the two- and three-way analysis of variance were used for the statistical treatment of the data.

Results

Psychomotor Skills

The results obtained on the 7th and 14th days did not differ from each other and they were therefore pooled for statistical handling.

Reactive skills. None of the drugs significantly modified cumulative choice reaction times and errors when compared with placebo. In the acute test DMD and O increased the reaction time of some subjects, but because of marked dispersion this effect was not statistically significant.

Coordinative skills. In coordination test I (driven at the fixed speed) ChL was the only drug which *per se* impaired coordination. In all groups receiving alcohol the coordination mistakes were significantly increased (fig. 1). In coordination test II (driven at free speed) the mistakes were increased in ChL + A and O + A groups. After the first single dose O prolonged the driving time significantly as also did the ChL + A group.

Attention. Of the single agents, only ChL tended to impair attention after a single dose, but this effect was not measurable after a week's treatment. It was also the only drug that showed significant interaction with alcohol. In the ChL + A group the number of correct responses was significantly lower than that in the placebo group.

Flicker fusion. This was the only test in which the morning and afternoon groups had different baseline performance: the threshold was significantly higher in the afternoon group. In the acute test DMD increased the threshold while O tended to lower it (fig. 2). After 1 or 2 weeks' treatment all drug effects disappeared.

Proprioception. After the first single dose of MO a significant impairment of proprioception was found. The ChL + A and MO + A groups showed poor results in later tests. The proprioceptions of hand and foot were equally deteriorated during these treatments.

Nystagmus. In the placebo groups, 13 % of the subjects had horizontal nystagmus whereas 47 % showed nystagmus after alcohol. The DMD + A, ChL + A, and MO + A groups had nystagmus even more often than those receiving alcohol only.

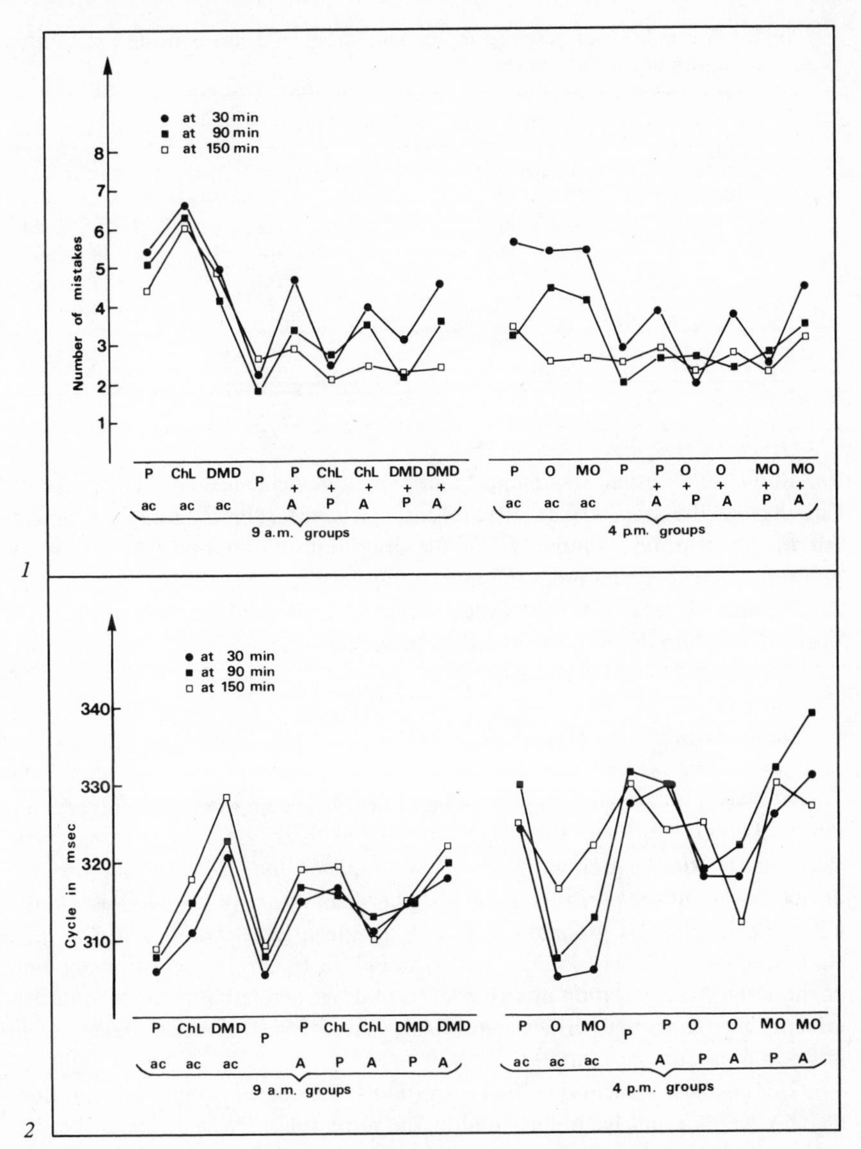

Fig. 1. Number of mistakes in coordination test I during the treatment with placebo (P), chlordiazepoxide lactam (ChL), desmethyldiazepam (DMD), oxazepam (O), or methyloxazepam (MO). Psychomotor skills are measured after the first dose of drugs (ac) and on the 7th and 14th day of treatment. The last-mentioned results are pooled for statistical handling. In these, alcoholic (A) or placebo (P) drink has been administered before the measurement of psychomotor skills. For further details see text.

Fig. 2. Discrimination of the fusion of flickering light during the treatment with different benzodiazepines with or without alcohol. For symbols see figure 1 and text.

Table I. Serum levels of the drugs on the 1st, 7th and 14th day of treatment, 3 h after the administration of last drug capsule

Drug	Serum levels of drugs, ng/ml (mean ± SD)		
	1st day	7th day	14th day
ChL	430 ± 50	2,550 ± 1,220	2,700 ± 1,500
DMD	82 ± 24	920 ± 440	1,220 ± 650
O	91 ± 75	270 ± 146	219 ± 111
MO	270 ± 120	710 ± 350	770 ± 490

For symbols see figure 1.

Other Assessments

In the acute test all treatments tended to improve the feeling of performance, ChL having the least effect. In subacute tests the subjects receiving alcohol felt their performance impaired. Of the drugs, only O improved the feeling of performance whereas all other treatments impaired it.

Pharmacokinetics. All the drugs except O tended to accumulate up to 2 weeks. The serum levels of the drugs are presented in table I.

Discussion and Conclusions

We have previously demonstrated a strong diazepam interaction on psychomotor skills (6, 8). Since the effects on these skills of the present compounds, alone and in combination with alcohol, were milder, the effects of diazepam are obviously due to the parent compound rather than its active metabolites. On the other hand, it seems probable that ChL significantly contributes to the mild chlordiazepoxide-alcohol interaction (9) found during prolonged administration of the drug. No correlation of serum levels of drugs and impairment of skills was evident suggesting that critical punishable serum levels of benzodiazepines in drivers are so far inappropriate.

This conclusion seems justified even though the parent compounds and their metabolites have not been compared in the same trial . The subjects used in the present study were of the same age, of a comparable educational level, and they came from the same living district as the subjects in the earlier Finnish studies (6–9). Further, the laboratory tests were essentially the same yet some extra tests causing minimum stress were included in the present trial. However, some reservations are necessary when interpreting the results:

(1) The time of day has not significantly affected one's reactive, coordinative and attentive skills in our laboratory conditions (6). This fact was confirmed

in the present experiment. The sensitivity of the flicker fusion to circadian effects was the only significant daytime-dependent variable measured in the present study. This finding does not exclude a possibility of a significant interaction of the drug effect and the time of the day. Accordingly, the results from the groups tested in the morning (ChL and DMD) may not be fully comparable with those of the groups O and MO tested in the afternoon.

(2) In previous acute studies in our laboratory (6, 7) the subjects were tested after a short warm-up period on the test apparatuses, whereas in the present study the subjects were well trained. This difference might be significant, particularly concerning the benzodiazepine interactions with alcohol. Our group has recently demonstrated this drug combination to be harmful on short-term memory and learning. However, in recent subacute studies like the present one, diazepam 5 mg t.i.d. strongly enhances the effect of alcohol (8). The only impairment measured after diazepam *per se* was a tendency not to compensate one's impairment in coordination test II slow driving.

The drug doses used in the present study gave twice as high serum levels of DMD and even much higher levels of ChL as compared with those observed in the studies on the parent drugs (8, 9). We have shown that a single oral dose of DMD greatly elevates its own serum levels after a second dose given 2 weeks later (5).

ChL proved the agent most harmful on attention and proprioception in combination with alcohol, and it might explain the observed chlordiazepoxide-alcohol interaction in a previous subacute study (9). Nystagmus proved very sensitive to the effect of alcohol, but it was positive in many subjects on placebo as well. Thus, unlike suggested earlier (2), it may be an inappropriate measure to be used for the detection of the alcohol-induced impairment of driving skills.

Summary

The interaction of the main metabolites of diazepam and chlordiazepoxide with alcohol was measured in two double-blind crossover subacute experiments on 40 healthy young volunteers. The drugs were administered for 2 weeks each. The variables measured were choice reaction time and accuracy, eye-hand coordination, divided attention, flicker fusion, proprioception, and nystagmus.

ChL, MO and O significantly enhanced the alcohol-induced impairment of psychomotor skills whereas DMD did so only exceptionally on some subjects in the choice reaction test. It is concluded that the diazepam-alcohol interaction on psychomotor skills is mainly due to the parent compound. No correlations between the serum levels of the agents and the changes of performance were found.

References

1 *Greenblatt, D.J. and Shader, R.I.:* Drug therapy: benzodiazepines. New Engl. J. Med. *291:* 1011–1015 (1974).

2 *Goldberg, L.:* Quantitative studies on alcohol tolerance in man. Acta physiol. scand. *5:* suppl. 16, pp. 1–128 (1943).

3 *Koechlin, B.A. and Darćonte, L.:* Determination of chlordiazepoxide and of a metabolite of lactam character in plasma of humans, dogs and rats by a specific spectrofluorometric micromethod. Analyt. Biochem. *5:* 195 (1963).

4 *Korttila, K.; Häkkinen, S., and Linnoila, M.:* Impairment of psychomotor skills by bupivacaine and etidocaine. Acta anaesth. scand. (in press).

5 *Korttila, K.; Mattila, M.J., and Linnoila, M.:* Saturation of tissues with *N*-desmethyldiazepam as a cause for elevated serum levels of this metabolite after repeated administration of diazepam. Acta pharmac. tox. *36:* (1975).

6 *Linnoila, M. and Mattila, M.J.:* Drug interaction on psychomotor skills related to driving: diazepam and alcohol. Eur. J. clin. Pharmacol. *5:* 186–194 (1973).

7 *Linnoila, M.:* Drug effects on psychomotor skills related to driving: interaction of atropine, glycopyrrhonium and alcohol. Eur. J. clin. Pharmacol. *6:* 107–112 (1973).

8 *Linnoila, M.; Saario, I., and Mäki, M.:* The effect of two weeks' treatment with diazepam and lithium alone or in combination with alcohol, on psychomotor skills related to driving. Eur. J. clin. Pharmacol. *7:* 337–342 (1974).

9 *Linnoila, M.; Saario, I., and Mäki, M.:* The effect of two weeks' treatment with chlordiazepoxide and flupenthixole, alone or in combination with alcohol, on psychomotor skills related to driving. Arzneimittel-Forsch. (in press).

10 *Vessman, J.; Freij, G., and Strömberg, S.:* Determination of oxazepam in serum and urine by electron capture gas chromatography. Acta pharm. suecica *9:* 447–456 (1972).

11 *Zingales, I.A.:* Diazepam metabolism during chronic medication: Unbound fraction in plasma, erythrocytes and urine. J. Chromat. *75:* 55 (1973).

Dr. *E.S. Palva,* Department of Pharmacology, University of Helsinki, *Helsinki* (Finland)

Mod. Probl. Pharmacopsych., vol. 11, pp. 85–90 (Karger, Basel 1976)

Two Weeks' Treatment with Chlorpromazine, Thioridazine, Sulpiride, or Bromazepam: Actions and Interactions with Alcohol on Psychomotor Skills Related to Driving

T. Seppälä, I. Saario and M.J. Mattila

Department of Pharmacology, University of Helsinki, Helsinki

In acute experiments many psychotropic drugs impair psychomotor skills to varying extents as demonstrated by different laboratory test combinations (1, 2). Subacute or long-term effects on the same parameters are not as well known since these trials are laborious and good cooperation is required. We have conducted two double-blind crossover trials, similar to each other, in order to assess the effects of 2 weeks' treatment with thioridazine (T), chlorpromazine (C), sulpiride (S) or bromazepam (B) on psychomotor skills. The interaction with alcohol (A) was also included to the study.

C and T are classic phenothiazines, while S is a new drug with antipsychotic and antidepressant properties, and it chemically differs from the traditional antipsychotics. The fourth drug in the study, B, is a benzodiazepine derivative. We chose the doses of phenothiazines to resemble the low doses used for outpatients as antianxiety agents.

Material and Methods

Both trials were double-blind and crossover, the consecutive treatments without washout being given in random order. The doses of C and T were 10 mg t.i.d. daily for the first 7 days and 20 mg t.i.d. for the next 7 days, while the doses of S 50 mg t.i.d. and B 6 mg t.i.d. were fixed over the whole period of 2 weeks. The drugs were given in identical gelatine capsules.

In the first trial we had 20 healthy, paid student volunteers and in the second trial 17. The criteria applied for choosing subjects were the same as reported earlier from this laboratory (1). The most important difference for acute studies was that all subjects were well trained on the test apparatuses before starting the study. The actual test procedure took place on the 7th and 14th days of each treatment. At every session half of the subjects received 0.5 g/kg of ethyl A as cooled bitter, and another half received non-alcoholic bitter

Table I. The laboratory test pattern used for screening the actions and interactions of drugs and alcohol

Parameter	Choice reaction test	Coordination tests I and II	Attention test	Flicker fusion test
Duration	48 sec	I 30 sec II 30–90 sec	10 min	–
Number of stimuli	32	–	840	–
Variables measured	reaction time number of mistakes	number of mistakes mistake percentage driving time (II)	number of responses number of correct responses	critical flicker fusion frequency
Factors measured	response orientation information retrieval from memory	control precision multilimb coordination rate control	information-processing capacity divided attention	central visual processes

of an equal volume. On every test day, all tasks were done three times, 30, 90 and 150 min after the intake of capsule and drink. The test battery and the main characteristics of the individual tests are presented in table I. They have been discussed in details previously (1).

At every test time, the subjects filled in a questionnaire where they were asked for the nature of their current treatment and an estimation about their performance. The data were treated according to the two- and three-way analyses of variance. The scores after different treatments were also compared with each other by means of Student's t test.

Results

The results from the two trials are presented separately. However, they are comparable with each other because the trial designs were identical and the test apparatuses the same in both trials. There were also parallel results after placebo (P) and A in both trials, and reproducibility of the tests has also earlier proved good.

In the first trial the active agents were C and S. As regards to the reactive skills, drugs alone had no significant effects on the 7th day when compared with

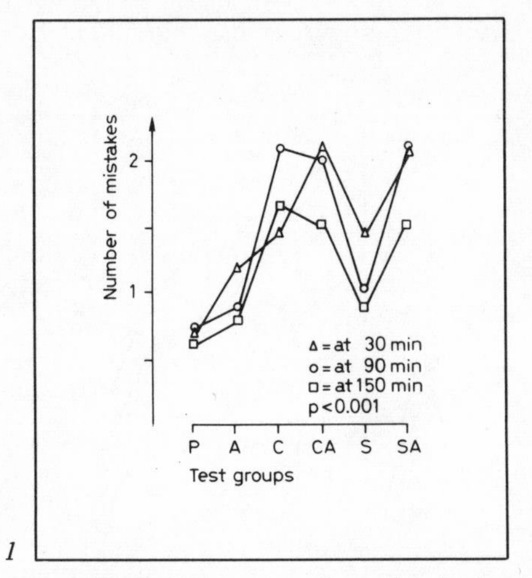

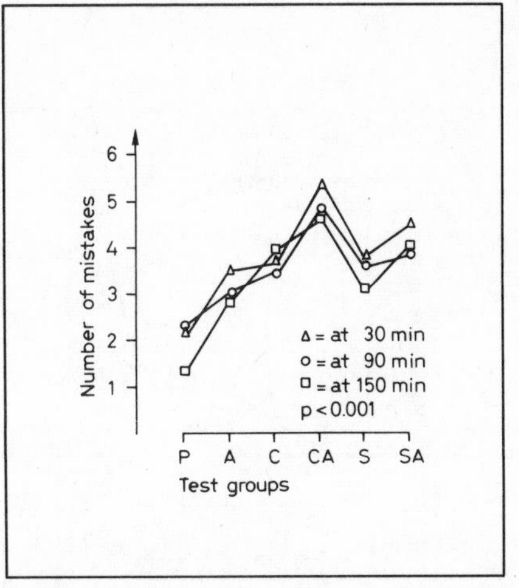

1 2

Fig. 1. The number of mistakes in the choice reaction test on the 14th day of the first trial with placebo (P), chlorpromazine (C) and sulpiride (S). A, CA and SA refer to the administration of alcohol (A) with placebo, chlorpromazine, or sulpiride, respectively. $p < 0.001$ for drug effect according to the three-way analysis of variance.

Fig. 2. The number of mistakes in coordination test II in the first trial. For symbols see figure 1. $p < 0.001$ for drug effect according to the two-way analysis of variance.

P. On the 14th day C impaired reaction accuracy, especially 90 and 150 min after the drug intake ($p < 0.05$) (fig. 1). Combinations of either drug with alcohol prolonged cumulative reaction times and increased the number of mistakes, most remarkable alterations being in the C + A group (fig. 1). The alterations in the A and S groups were not statistically significant when compared with the P group by means of Student's t test.

As regards to the coordinative skills in coordination test I (driven at the fixed speed) the results with C were different on the 7th and 14th days, so that the number of mistakes and the mistake percentage were elevated only on the 14th day. On this test day the performance of this group deteriorated towards the end of the session. In the C + A group both parameters measured were significantly increased ($p < 0.05$) on both the 7th and 14th days. After S, alone or with A, no impairment of skills were observed in this test.

Concerning coordination test II (driven at free speed) there were no statistically significant differences between the results observed on the 7th and 14th days. Therefore, the results at these two sessions were pooled and handled according to the two-factor design, the factors influencing the final analysis being the test time and the treatment. Figure 2 shows the number of mistakes in coordination test II. When compared with the corresponding P values, the trend of impairment was significant in the C + A group ($p < 0.01$) and in the S + A

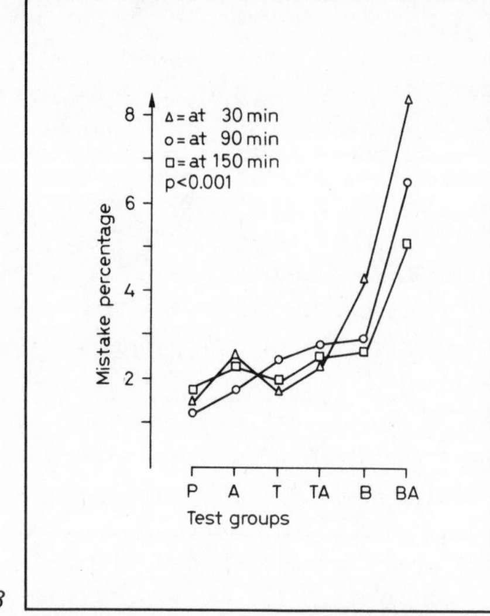

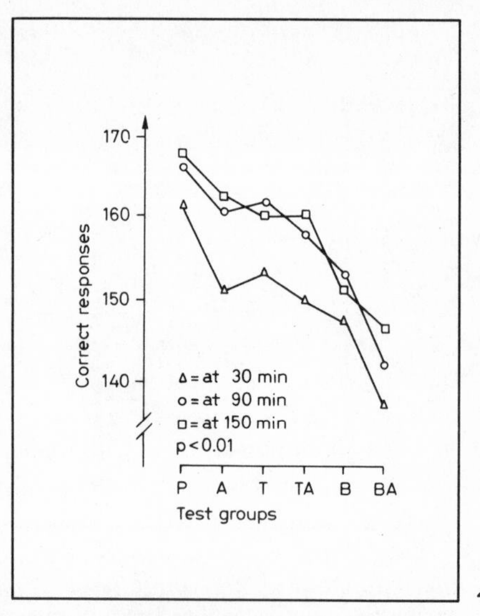

Fig. 3. The mistake percentage in coordination test II in the second trial with placebo (P), thioridazine (T) and bromazepam (B). A, TA and BA refer the administration of alcohol with placebo, thioridazine, or bromazepam, respectively. $p < 0.01$ for drug effect according to the two-way analysis of variance.

Fig. 4. Total numbers of correct responses in the attention test in the second trial. For symbols see figure 3. $p < 0.01$ for drug effect according to the two-way analysis of variance.

group ($p < 0.05$) at every test time, and also in the C group at 150 min ($p < 0.05$). The C + A subjects drove significantly slower than other groups in this test, but they could not compensate their mistakes in this way. Attention and ability to discriminate the fusion of flickering light were not affected after any treatment in the first trial.

In the second trial the active agents were T and B. Because the test week had no influence on any variable recorded, the data from the two test days are handled together. B was the only agent which increased the number of mistakes ($p < 0.05$) in the choice reaction test.

Recordings from both coordination tests were parallel. Figure 3 shows the mistake percentages in coordination test II in different groups. The mistake percentage was very high in the B + A group decreasing gradually up to 150 min. The mistake percentage was also significantly ($p < 0.05$) elevated in the B group in a 30-min test. T did not differ from P.

In the attention test the total numbers of responses were not significantly altered, but the numbers of correct responses were ($p < 0.05$) smaller in the B and B + A groups than in the P group (fig. 4). This decrease resulted from diminished numbers of correct responses on peripheral dials in both groups. The ability to discriminate the fusion of flickering light also remained unaltered after treatments of the second trial. As a whole, the subjective assessments tallied well with the objective data after every treatment in both trials.

Discussion

Traffic safety is a complex matter and the present study touches only one angle from which the problem can be approached. However, it seems unlikely that the effects recorded should be without importance for estimation of driving ability. S and T alone seem to be inactive in the doses used in the subacute study, while C after 2 weeks' treatment may be harmful for driving. B caused the most effects in this study, which may partly result from the dose which has appeared relatively higher than the doses of other drugs. Even then one may say that antipsychotics exert an additive and B a true synergistic interaction with A. In connection to this experiment, the effects of these drugs on memory and learning were studied using the same subjects. The effects of drugs on learning were in good correlation with that on psychomotor performance. B impaired learning and the combined effect of B and A was very deleterious.

Summary

Subacute effects of C, T, S, or B, alone or in combination with A, were tested against P in two double-blind crossover trials with 37 healthy students. The drugs were given in capsules t.i.d. for 2 weeks each and the psychomotor performance (choice reaction, coordination, attention) was measured on the 7th and 14th days of treatment. At each session the subjects swallowed a capsule together with 0.5 g/kg of A or P drink, and the measurements were done 30, 90, and 150 min thereafter.

T alone did not differ from placebo in the doses used (10 mg t.i.d. for 7 days and 20 mg t.i.d. for the next 7 days). After C (dosing as above) and S (50 mg t.i.d.) both reactive and coordinative skills were slightly impaired. B (6 mg t.i.d.) clearly impaired both reactive skills and attention.

T + A had no major combined effect on skills while C interacted with A resulting in impaired reactive and coordinative skills. After C + A the subjects were unable of compensating their coordination mistakes by slow driving. The interaction of S with A was mild, whereas B + A strongly impaired coordination and divided attention. No alterations were recorded in flicker fusion after any treatment.

The results suggest that low doses of neuroleptics impair psychomotor skills less than benzodiazepines do, but therapeutically equipotent doses must be assessed in order to apply the results into practice.

References

1 *Linnoila, M. and Mattila, M.J.:* Drug interaction on psychomotor skills related to driving: diazepam and alcohol. Eur. J. clin. Pharmacol. *5:* 186–194 (1973).
2 *Milner, G. and Landauer, A.:* Alcohol, thioridazine and chlorpromazine effects on skills related to driving behaviour. Br. J. Psychiat. *118:* 351–352 (1971).

Dr. *T. Seppälä,* Department of Pharmacology, University of Helsinki, *Helsinki* (Finland)

Mod. Probl. Pharmacopsych., vol. 11, pp. 91–98 (Karger, Basel 1976)

Minor Outpatient Anaesthesia and Driving

Kari Korttila

Departments of Anaesthesia and Pharmacology, University of Helsinki, Helsinki

Rapidly increasing hospital expenses have produced a common trend towards outpatient anaesthesia and sedation. Consequently, a rapid recovery after anaesthesia and the knowledge of the duration of the harmful effects of anaesthetic agents on human psychomotor performance is essential. The question 'for how long should patients not drive after outpatient anaesthesia or sedation' has been discussed several times and is of increasing interest, since the number of cars and people participating in traffic continues to increase.

In a survey of 100 cases with outpatient anaesthesia, it was discovered that, except on clear instructions to the contrary, 31 patients went home unescorted by a responsible person. Out of 41 car owners 4 drove themselves home, and 30 drove within 24 h from the operation. A bus driver returned to his job on the same day (20).

Premedication, local anaesthetics and anaesthetics can impair patients' psychomotor skills for variable lengths of time. The aim of this communication is to present some data concerning the effects of anaesthetics and premedication on the psychomotor skills related to driving.

Assessment of Residual Effects of Anaesthetics

Attempts to subjectively or objectively measure recovery from anaesthesia have ranged from measurements of the ability of patients to open their eyes to their ability to drive a car. Various stages of recovery after anaesthesia, as well as tests for measuring them, are shown in table I. Various clinical tests have been used as a guideline to the evaluation of recovery. These are not, however, adequate indications of a patient's ability to react safely to his environment (14, 25).

Various simple paper and pencil tests have been used in assessing recovery after anaesthesia (1, 24). The advantage of these tests is that they can be used

Table I. Stages of recovery from anaesthesia and tests valid for their evaluation

Stage of recovery	Tests of recovery
Awakening	opening eyes answering
Immediate clinical recovery	sitting steady negative Rombergism and other clinical tests
Fit to go home[1] (= hospital stay)	Maddox-Wing paper and pencil tests single reaction time tests single coordination or attention tests flicker fusion
'Street fitness'[1]	flicker fusion psychomotor test batteries EEG
Complete psychomotor recovery[1] (= fit to drive)	carefully selected psychomotor test batteries driving simulators

[1] More than a single test is needed.

easily in clinical practice. When used alone, however, they do not seem to be sensitive enough to evaluate complete psychomotor recovery (14), and when used for assessing the length of hospital stay after anaesthesia they should be designed to measure both attention and coordination, or even reactive skills.

Doenicke et al. (2) have demonstrated that EEG measurements give valuable information about recovery times after anaesthesia. Drawbacks of the EEG are that it must be observed continuously throughout the recovery period and that it is sensitive to electrical interference. Measurements of ocular imbalance by Maddox-Wing (6) and the subjects' ability to discriminate the fusion of a flickering light have proved useful in evaluating recovery from anaesthesia or sedation. As a single task they cannot be considered to measure psychomotor recovery (19), since, when measured in a controlled way, the ability to see a flickering light seems to be of little or no clinical significance as regards driving ability (15).

Measurement of reaction time may be employed to determine how rapidly the patient is able to react to environmental stimuli. Reactions to acoustic or optical stimuli have been generally used, but impairment of reactive skills have proved to be greater and of longer duration when a choice reaction test is used (21). *Doenicke et al.* (2) have used a 'track-tracer' technique which consists of the tracking of a curving pathway. The same workers have also employed a psychophysiological test battery and found it suitable for measuring the late effects of anaesthetic agents.

Green et al. (4) and *Wilkinson* (26) have used a simple driving simulator in evaluating patients' recovery after anaesthesia, and *Korttila et al.* (13) with a complete driving simulator having other vehicles on the road. The value of driving simulators, reaction timers, and complex psychodiagnostic test batteries is that they provide reliable objective information about the degree and duration of the harmful effects of anaesthetic agents, but they are not suitable for use in everyday clinical practice.

Assessment of Driving Ability

Environmental factors, such as weather conditions, the quality of roads, and the amount of traffic, as well as the condition of the car, contribute to the possibility of a traffic accident. In most accidents, however, the driver himself causes the accident due to his attitude or skills. The latter can only be objectively measured. The psychomotor test battery used in evaluation of driving skills as well as the sufficiency of hospital stay after anaesthesia includes at least three types of tests: choice reaction, coordination, and divided attention tasks (19) already presented to the Symposium by previous speakers. Driving simulators should be as natural as possible and the less arbitrary the instructions are, the more reliable the results. Emergency situations are particularly effective in revealing drug effects.

'Real driving tasks' have usually consisted of skilled tracking and parking, which are not very reliable in predicting the effects of fatigue or drugs on real driving because they do not generally measure divided attention. A field test can be replaced by partial simulation (8). It has been suggested that in order to measure driving skills (19) one should be aware of exactly which psychomotor functions the tests measure. Further, one has to test several hundred subjects without any treatment in order to find out the correlation of the tests to traffic or occupational behaviour as well as their reliability and validity.

Effects of Anaesthetics and Adjuvants on Human Psychomotor Performance

Premedication and Analgesics

Orally taken anticholinergics may impair patients' psychomotor skills (18). Drowsiness and, in some instances, even slight ataxia have been noticed after parenteral doses of 0.03 mg/kg atropine. The small dose of approximately 0.5 mg atropine used in premedication to anaesthesia probably does not affect human psychomotor performance (13). 10 mg diazepam and 75 mg pethidine injected intramuscularly impaired psychomotor skills for at least 5–7 and 12 h, respectively (9), whereas no residual effects were noticed by a battery of psychological tests and EEG 2 h after intravenous injection of 0.1 mg or 8 h after 0.2 mg fentanyl to normal subjects (5). A role of narcotic antagonist in abolishing the residual effects of narcotics remains to be investigated. Of non-narcotic analgesics aspirin does not impair psychomotor skills whereas phenylbutazone and indomethacin enhance the alcohol effect.

Local Anaesthetics

Local anaesthetics are recommended as preferable agents for outpatient anaesthesia, mainly because it is believed that they do not affect the psychomotor performance of the patients. Delayed reaction times seen after local dental anaesthesia are believed to result from the stress because of the procedure and recommendations not to drive or operate machinery for an hour after dental procedures under local anaesthesia have been made (21, 22).

In higher doses, psychomotor performance was impaired for 1–1.5 h after 200 mg of plain lidocaine whereas 500 mg lidocaine with adrenaline did not impair psychomotor skills (10). Since prilocaine is less toxic to central nervous functions than lidocaine, prilocaine might, in this respect, be safer than lidocaine for outpatient use. We have found that the long-acting local anaesthetics, bupivacaine (1.3 mg/kg) and etidocaine (2.6 mg/kg), may impair psychomotor performance for at least 2 h if solutions without adrenaline are used (15). When no contraindications occur, their solutions with adrenaline are preferable to plain solutions in outpatient practice.

Psychotropic Drugs

Recent studies (5, 10) indicate that psychomotor skills are impaired for at least 6 h after 0.15 mg/kg of intravenous diazepam and 10 h after 0.45 mg/kg. The increase in the plasma levels of diazepam which seem to correlate with the recurrence of clinical sedation, is presumably due to the fact that the subjects had eaten a meal which resulted in elevated serum levels of diazepam.

When 7 subjects received 0.3 mg of diazepam intravenously and then a meal in either 3 or 7 h, an increase in post-prandial diazepam serum levels (20 and 50 %, respectively) coincided with impairment of coordinative and reactive skills at 3 h only (16). An explanation of this could be remobilization of diazepam from its storage site in the gastrointestinal tract (16).

A new benzodiazepine flunitrazepam has more prolonged residual effects than diazepam given in clinically comparable doses. Electroencephalographic sleep patterns have been demonstrated at 18 h and an impairment of performance on behavioural tests at 12 h after oral intake of 1 or 2 mg flunitrazepam. After 0.01 mg/kg flunitrazepam intravenously, eye-hand coordination was slightly impaired for as long as 6 h after the injection, and after 0.02 and 0.03 mg/kg the impairment still was significant at the last observation period 10 h after the injection (17).

After the administration of 5 mg droperidol, with or without fentanyl, the harmful residual effects were definitely more marked than those of diazepam and flunitrazepam. Psychomotor performance was severely impaired for at least 10 h after the administration (11).

Intravenous Anaesthetic Agents

Early studies on thiopentone and methohexitone suggested that recovery would occur in 1 or 2 h. However, later reports have demonstrated sleep patterns in an EEG 12 h after doses of 2 mg/kg methohexitone (3), and impaired test performance 8 h after doses of 500 mg thiopentone (13). Although awakening, immediate clinical recovery, and psychomotor recovery 1–4 h after anaesthesia is faster after methohexitone than after thiopentone, complete psychomotor recovery seems to take the same length of time with both agents in equianaesthetic doses. Impairment of driving skills for up to at least 8 h after doses of 6.0 mg/kg thiopentone and 2.0 mg/kg methohexitone was demonstrated using a driving simulator (13).

Doenicke et al. (2) found neither electroencephalographic sleep patterns 30 min or more after injections of 500–1,000 mg propanidid nor any deterioration of psychomotor performance 60 min after these injections. On the other hand, delayed reaction times to acoustic or optical stimuli 2 h after propanidid anaesthesia have been reported, as compared with the pretreatment results. With driving simulator no impairment of driving skills could be observed 2 h or more after the injection of 6.6 mg/kg propanidid (13).

The ocular imbalance assessed by Maddox-Wing may return to the pretreatment level 90 min after the injection of 50 μl/kg alphadione (6). The results of simulated driving indicate a late impairment of driving skills at 6 h after 85 μl/kg alphadione whereas the performance was similar to controls when measured at 2 and 4 h after injection (13). The latter agrees with clinical observations and tests which show that patients anaesthetised with alphadione recovered in 2 h, whereas those patients treated with methohexitone showed slower recovery (6).

Recovery from ketamine anaesthesia does not seem to be faster than from thiopentone anaesthesia (23). Although etomidate alone seems to be devoid of harmful late effects, the fact that it cannot be used as a sole anaesthetic without a rather heavy premedication or without combining it with narcotics (7) is a drawback in outpatient anaesthesia.

Inhalation Agents

Due to rapid recovery nitrous oxide offers an ideal supplementation for outpatient anaesthesia. In a recent study (*Korttila et al.*, to be published)

Table II. Outpatient anaesthesia: recommended minimal values for the length of hospital stay and the length of time patients should be advised against driving

Treatment	Hospital stay h	No driving h
Dental local anaesthesia dose variable	unnecessary	1
Plain lidocaine 200 mg i.m.	1	1–1.5
Lidocaine 500 mg with adrenaline i.m.	0.5	no limitations
Plain bupivacaine 1.3 mg/kg i.m.	2	2–4
Plain etidocaine 2.6 mg/kg i.m.	2	2–4
Diazepam 10 mg i.m.	1	7
Pethidine 75 mg i.m. or i.v.[1]	2–3	24
Fentanyl 0.1 mg i.v.	1–2	2
Fentanyl 0.2 mg i.v.	2	8
Diazepam i.v. 0.15 mg/kg	2	8
0.30 mg/kg[2]	2–3	10
0.45 mg/kg[2]	3–4	10
Diazepam 0.15 mg/kg plus[1] Pethidine 1mg/kg i.v.	2–3	10–24
Flunitrazepam 0.01 mg/kg i.v.	2	8
Flunitrazepam 0.02–0.03 mg/kg i.v.[1]	3–4	24
Flunitrazepam 0.015 mg/kg plus[1] Pethidine 1 mg/kg i.v.	3	24
Droperidol 5 mg i.v.[1]	4–6	24
Droperidol 5 mg plus[2] Fentanyl 0.2 mg i.v.	4–6	24
Thiopentone 6.0 mg/kg i.v.[3]	3	24
Methohexitone 2.0 mg/kg i.v.[3]	2	24
Propanidid 6.6 mg/kg i.v.	1–2	3–4
Alphadione 85 μl/kg i.v.	2	8
Halothane-N_2O-O_2 (5–10 min)	2	7
Enflurane-N_2O-O_2 (5–10 min)	2	7

[1] Treatment should be avoided in outpatient practice.
[2] Recurrence of impaired performance should be taken into consideration, if food is had within less than 5 h.
[3] Other treatment preferable in outpatient practice.

psychomotor performances remained significantly worse than in a control group for 5 h after brief halothane and enflurane anaesthesia, and driving skills in driving simulator 4.5 h after halothane were worse than in control subjects. It seems that recovery after cyclopropane is more rapid than after thipentone, whereas methoxyflurane has longer-lasting actions.

Conclusions

It seems clear that patients should always be escorted when leaving the hospital after minor outpatient anaesthesia. The length of hospital stay should be based on the patients' test performance as well as on the supposed effects of the drugs on psychomotor performance. Recommendations not to drive should be based on documented objective knowledge of the late effects of the drugs used, as well as on the extent of the impairment of performance when assessed at the hospital.

On the basis of our own results and of the information found in literature, recommendations have been made about the lengths of time of hospital stay and of time that they should be advised to refrain from driving after outpatient anaesthesia (table II).

References

1 *Dixon, R.A. and Thornton, J.A.:* Tests of recovery from anaesthesia and sedation: intravenous diazepam in dentistry. Br. J. Anaesth. *45:* 207–215 (1973).

2 *Doenicke, A.; Kugler, J., and Laub, M.:* Evaluation of recovery and 'street fitness' by EEG and psychodiagnostic tests after anaesthesia. Can. Anaesth. Soc. J. *14:* 567–583 (1967).

3 *Doenicke, A.; Kugler, J.; Spann, W., et al.:* Hirnfunktion und psychodiagnostische Untersuchungen nach intravenösen Kurznarkosen und Alkoholbelastungen. Anaesthesist *15:* 349–355 (1966).

4 *Green, R.; Long, H.A.; Elliot, C.J.R., and Howells, T.H.:* A method of studying recovery after anaesthesia. Anaesthesia *18:* 189–200 (1963).

5 *Ghoneim, M.M. and Metwaldt, S.P.:* The effect of diazepam or fentanyl on mental, psychomotor and electroencephalographic functions and their rate of recovery. 6th Int. Congr. of Pharmacology, abstract. Psychopharmacologia *44:* 61–66 (1975).

6 *Hannington-Kiff, J.G.:* Comparative recovery rates following induction of anaesthesia with althesin and methohexitone in outpatients. Post-grad. med. J. *48:* suppl. 12, pp. 116–119 (1972).

7 *Hempelman, G.; Hempelmann, W.; Prepenbrock, S., et al.:* The influence of etomidate on blood gases and haemodynamics in patients with myocardial disease. Anaesthesist *23:* 423–429 (1974).

8 *Kielholz, P.; Goldberg, L.; Hobi, V. und Reggiani, G.:* Teilstimulation zur Prüfung der Beeinträchtigung der Fahrtüchtigkeit unter Alkohol. Schweiz. med. Wschr. *101:* 1725–1731 (1971).

9 *Korttila, K. and Linnoila, M.:* Psychomotor skills related to driving after intramuscular administration of diazepam and meperidine. Anesthesiology *42:* 685–691 (1975).

10 *Korttila, K.:* Psychomotor skills related to driving after intramuscular lidocaine. Acta anaesth. scand. *18:* 290–296 (1974).

11 *Korttila, K. and Linnoila, M.:* Skills related to driving after intravenous diazepam, flunitrazepam or droperidol. Br. J. Anaesth. *46:* 961–969 (1974).

12 *Korttila, K. and Linnoila, M.:* Recovery and skills related to driving after intravenous sedation: dose-response relationship with diazepam. Br. J. Anaesth. *47:* 457–463 (1975).

13 *Korttila, K.; Linnoila, M.; Ertama, P., and Häkkinen, S.:* Recovery and simulated driving after intravenous anesthesia with thiopental methohexital propanidid or alphadione. Anesthesiology *43:* 291–299 (1975).

14 *Korttila, K.:* Recovery after intravenous sedation. A comparison of clinical and paper and pencil tests used in assessing late effects of diazepam. Anaesthesia (in press, 1976).

15 *Korttila, K.; Häkkinen, S., and Linnoila, M.:* Side effects and skills related to driving after intramuscular administration of bupivacaine and etidocaine. Acta anaesth. scand. *19:* 384–391 (1975).

16 *Korttila, K.; Mattila, M.J., and Linnoila, M.:* Prolonged recovery after diazepam sedation. The influence of food, charcoal ingestion and injection rate on the effects of intravenous diazepam. Br. J. Anaesth. (in press, 1976).

17 *Korttila, K. and Linnoila, M.:* Amnesic action and skills related to driving after intravenous flunitrazepam. Acta anaesth. scand. (in press, 1976).

18 *Linnoila, M.:* Drug effects on psychomotor skills related to driving: interaction of atropine, glykopyrronium and alcohol. Eur. J. clin. Pharmacol. *6:* 107–112 (1973).

19 *Linnoila, M.; Saario, I.; Seppälä, T., et al.:* Methods used for evaluation of the combined effects of alcohol and drugs on humans; in *Morselli, Garattini and Cohen* Drug interactions, pp. 319–325 (Raven Press, New York 1974).

20 *Ogg, T.W.:* An assessment of postoperative outpatient cases. Br. med. J. *iv:* 573–575 (1972).

21 *Tetsch, P.:* Reaktionszeitmessungen bei zahnärztlich-chirurgischen Eingriffen in Analgosedierung. Dt. zahnärztl. Ztg *28:* 618–622 (1973).

22 *Tetsch, P.; Machtens, E. und Voss, M.:* Reaktionszeitsmessungen bei operativen Eingriffen in örtlicher Schmerzausschaltung. Schweiz. Mschr. Zahnheilk. *82:* 229–306 (1972).

23 *Thompson, G.E.; Remington, J.M., and Millman, B.S.:* Experiences with outpatient dental anaesthesia. Anesth. Analg. curr. Res. *52:* 881–887 (1974).

24 *Trieger, N.; Newman, M.G., and Miller, J.G.:* An objective measure of recovery. Anesth. Prog. *16:* 4–9 (1969).

25 *Vickers, M.D.:* The measurement of recovery from anaesthesia. Br. J. Anaesth. *37:* 296–302 (1965).

26 *Wilkinson, B.M.:* Driving ability and reaction times following intravenous anaesthesia. N.Z. dent. J. *61:* 21–26 (1965).

Dr. *K. Korttila,* Departments of Anaesthesia and Pharmacology, University of Helsinki, *Helsinki* (Finland)

Closing Remarks

This symposium has been both stimulative and useful in both pointing out the main problems of present research and in outlining important future topics. One topic of special impetus which has come up is the need for *international cooperation* in the field of traffic research. This is particularly true in epidemiological studies, but also laboratory methods measuring psychomotor skills require standardization, in case in the future certain drug groups are pretested in this respect before marketing them.

As to *drinking and driving,* the main problem seem to be young drivers consuming fair amounts of alcohol before driving. They represent the main toll of fatal traffic accidents on the highways during weekend nights. This pattern of drinking and driving becomes evident in all western countries where traffic fatalities are adequately examined. It is revealed in the roadside surveys as well. Our countermeasures seem to be inadequate because the number of drinking drivers is increasing. The only effective large-scale countermeasure which has been accidentally applied is the restriction of availability of alcohol. The strikes of the state alcohol monopolies in Sweden and Finland have dramatically reduced alcohol-related fatalities in both countries. A high taxation of alcoholic beverages and adjustment of the taxes according to inflation may yield similar though less evident results. Severe penalties may affect social drinkers, but they are probably without effect on alcoholics. The agreement on the methods of roadside surveys reached in the IDBRA meeting in Paris in 1974 will hopefully lead to comparable designs of these studies in different countries, and thus provide an opportunity for large-scale studies concerning the efficacy of different methods of intervention.

As to the *traffic accident risk caused by drugs* other than alcohol, the argument over whether or not the drugs cause any increase of accident risk in their users is still valid, though less so than before this meeting. Recent results of a Norwegian hospital study demonstrated diazepam in the serum of 18 % of injured traffic accident victims who were hospitalized. Our questionnaire study demonstrated that the traffic exposure balanced accident involvement during the 2 years previous to the study was more than twice as high among medicated psychiatric outpatients than among non-medicated psychiatric patients and age and living district matched controls. Thus it seems that psychotropic drugs may increase traffic accident risk. However, even among the psychotropic drugs clear-cut differences exist as to their potency to impair skills related to driving.

The main *methodological difficulty* in conducting roadside surveys concerning drug use among drivers is sampling of blood. This is necessary in order to analyze adequately the quality and quantity of the drug used before driving. While the percentage of drivers cooperating in roadside surveys concerning alcohol is well above 90 %, that in recent American studies concerning drugs has been around 70 %. It is clear that epidemiological studies with a drop-out rate of about one third of the potential sample are meaningless. Particularly, because one has reason to believe that the drop out is selective, i.e. drivers who have been taking drugs may be reluctant to participate. Once the blood sample is drawn, an analysis with gas chromatography mass spectrometry is capable of measuring most drugs.

Because the epidemiology of drug use is such that tranquilizers and sedative hypnotics are mainly used by elderly and middle-aged females, these drugs are not expected to be overrepresented among fatal highway accidents. Thus, studies concerning the victims of fatal traffic accidents should not be expected to reveal a significant overrepresentation of the users of these drugs. They generally are not driving at high speeds on highways at night. On the other hand, *hospital studies* concerning the victims of personal injury accidents occurring during the day-time and in the evenings, as well as random sampling among drivers involved in property damage accidents should reveal the real impact of these drugs on traffic accident risk if the control groups are adequately matched.

In the *laboratory* the effects of drugs on perception, information sampling, choice reaction times, eye-hand coordination, divided attention, proprioception, and vestibular function can be easily measured. Information about the drug action on mood and particularly traffic attitudes is less reliable. Changes in skills after the drugs can be correlated with the blood levels of the drug and its main active metabolites. Both healthy volunteers and volunteer patients should be used as subjects. The main pitfall of laboratory studies has been the lack of patients as subjects. Thus the application of their results in real-life situations is in dispute. However, the drug consumption statistics of the industrial countries reveal that a large number of relatively healthy individuals are on drugs. This is particularly true for the minor tranquilizers.

The main duty of the pharmaceutical industry is naturally to develop specific and effective drugs with feasible side effects. Some difficulties encountered are well elucidated by an example concerning anxiolytics. There are two well done Scandinavian studies, one Finnish and one Swedish, demonstrating that the anxiolytic agent in patients taking diazepam is the parent compound. There is also a well-conducted British study which claims that the active ingredient as to anxiolysis is the main metabolite of diazepam, *N*-desmethyldiazepam. In our laboratory we have demonstrated that *N*-desmethyldiazepam is less deleterious to driving skills in healthy volunteers than diazepam. This is especially true as to the interaction of these drugs with alcohol. However, it is hard to recommend the use of *N*-desmethyldiazepam as long as there is dispute concerning its efficacy.

To sum up the recommendations one has picked up in this conference for future research, I think that at the moment different kinds of interventions into the drinking and driving problem should be tried and that their value should be measured by roadside surveys coordinated internationally. Measures restricting alcohol availability in a selective manner, e.g. during weekend nights, may turn out to be useful. The involvement of drug users in traffic accidents should be investigated by internationally cooperated hospital studies. The laboratory studies should use both patients and healthy volunteers as subjects and multiple measurements should be used to discover the effects of drugs on the many functions necessary for proper driving. In the case of tranquilizers and sedatives, the drug-alcohol interaction has to be studied as well.

M. Linnoila, Helsinki

Subject Index